Visible Light Communications (Second Edition)

Vehicular applications

Online at: https://doi.org/10.1088/978-0-7503-6049-4

IOP Series in Emerging Technologies in Optics and Photonics

Series Editor

R Barry Johnson, a Senior Research Professor at Alabama A&M University, has been involved for over 50 years in lens design, optical systems design, electro-optical systems engineering, and photonics. He has been a faculty member at three academic institutions engaged in optics education and research, has been employed by a number of companies, and has provided consulting services.

Dr Johnson is an IOP Fellow, an SPIE Fellow and Life Member, an OSA Fellow, and was the 1987 President of SPIE. He serves on the editorial board of *Infrared Physics & Technology* and *Advances in Optical Technologies*. Dr Johnson has been awarded many patents, has published numerous papers and several books and book chapters, and was awarded the 2012 OSA/SPIE Joseph W Goodman Book Writing Award for Lens Design Fundamentals (second edition). He is a perennial co-chair of the annual SPIE Current Developments in Lens Design and Optical Engineering Conference.

Foreword

Until the 1960s the field of optics was primarily concentrated in the classical areas of photography, cameras, binoculars, telescopes, spectrometers, colorimeters, radiometers, etc. In the late 1960s optics began to blossom with the advent of new types of infrared detector, liquid crystal display (LCDs), light emitting diode (LEDs), charge coupled device (CCDs), laser, holography, and fiber optics along with new optical materials, advances in optical and mechanical fabrication, new optical design programs, and many more technologies. With the development of the LED, LCD, CCD, and other electro-optical devices, the term ‘photonics’ came into vogue in the 1980s to describe the science of using light in the development of new technologies and the operation of a myriad of applications. Today optics and photonics are truly pervasive throughout society and new technologies are continuing to emerge. The objective of this series is to provide students, researchers, and those who enjoy self-education with a wide-ranging collection of books, each of which focuses on a topic relevant to the technologies and applications of optics and photonics. These books will provide knowledge to prepare the reader to be better able to participate in these exciting areas now and in the future. The title of this series is *Emerging Technologies in Optics and Photonics*, in which ‘emerging’ is taken to mean ‘coming into existence’, ‘coming into maturity’, and ‘coming into prominence’. IOP Publishing and I hope that you will find this series of significant value to you and your career.

A full list of titles published in this series can be found here: https://iopscience.iop.org/bookListInfo/emerging-technologies-in-optics-and-photonics.

Visible Light Communications (Second Edition)

Vehicular applications

Xavier Fernando
Department of Electrical and Computer Engineering, Ryerson Communications Lab, Toronto, Ontario, Canada

Hasan Farahneh
Department of Electrical Engineering, University of Jordan, Amman, Jordan

IOP Publishing, Bristol, UK

ISBN 978-0-7503-6049-4 (ebook)
ISBN 978-0-7503-6047-0 (print)
ISBN 978-0-7503-6050-0 (myPrint)
ISBN 978-0-7503-6048-7 (mobi)

DOI 10.1088/978-0-7503-6049-4

Version: 20241201

IOP ebooks

British Library Cataloguing-in-Publication Data: A catalogue record for this book is available from the British Library.

Published by IOP Publishing, wholly owned by The Institute of Physics, London

IOP Publishing, No.2 The Distillery, Glassfields, Avon Street, Bristol, BS2 0GR, UK

US Office: IOP Publishing, Inc., 190 North Independence Mall West, Suite 601, Philadelphia, PA 19106, USA

Contents

Preface ix

Acknowledgements xiii

Author biographies xiv

List of acronyms xv

1 Introduction **1-1**

1.1 Optical wireless communication 1-1
1.1.1 Cellular V2X for intelligent transportation systems 1-4
1.1.2 Visible light technology in V2X communications 1-6
1.2 VLC for intelligent transportation 1-9
1.2.1 VLC in the ITS/state of the art 1-10
1.3 AI-based V2X VLC communications 1-12
1.4 The autonomous vehicle 1-14
1.4.1 The role of visible light in AVs 1-14
1.5 VLC/OWC communications to flying objects 1-15
1.5.1 VLC/OWC for UAV-enhanced IoV 1-16
1.5.2 VLC/OWC for UAV-LEO satellite communications 1-16
1.5.3 AI-enhanced UAV-satellite communications 1-17
1.6 VLC V2X research directions 1-18
1.7 Chapter summary 1-19
References 1-19

2 Nuts and bolts of V2X VLC systems **2-1**

2.1 VLC basics 2-1
2.1.1 Visible light terminology 2-1
2.2 VLC emitter 2-2
2.2.1 LED or laser transmitter 2-4
2.3 VLC receiver 2-5
2.3.1 Image sensors as VLC receivers 2-9
2.4 VLC channel 2-11
2.5 Comparison of VLC and RF-based V2X communications 2-12
2.6 VLC advantages and drawbacks 2-13
2.7 Safety concerns of OWC systems 2-13

2.8 An example collision avoidance system 2-14
2.8.1 Proposed system 2-15
2.8.2 Design of algorithm 2-17
2.8.3 Simulation and results 2-19
2.9 Chapter summary 2-24
References 2-25

3 Impacts of directional propagation 3-1
3.1 Traffic light-to-vehicle communication scenario 3-1
3.1.1 Novel potential improvements 3-7
3.2 Shadowing effects 3-8
3.2.1 Bimodal Gaussian reception 3-9
3.2.2 Channel impulse response with shadowing 3-11
3.2.3 Power spectral density of the received signal 3-13
3.2.4 Error probabilities with and without shadowing 3-14
3.3 Chapter summary 3-18
References 3-18

4 Channel modeling for V2V-VLC system 4-1
4.1 Channel modeling for VLC 4-1
4.1.1 Intensity modulation and direct detection (IM/DD) 4-2
4.2 Indoor VLC channel models 4-3
4.2.1 LoS propagation model 4-3
4.2.2 NLoS propagation model 4-4
4.2.3 Ceiling bounce model 4-5
4.2.4 Spherical model 4-6
4.3 Outdoor V2V-VLC channel modeling 4-6
4.3.1 History of channel modeling of vehicular communication 4-6
4.3.2 2×2 MIMO V2V-VLC system model 4-7
4.3.3 The V2V-VLC system considerations 4-7
4.3.4 System description 4-8
4.4 Channel impulse response and transfer function 4-10
4.4.1 Impulse response and transfer function of the LoS component 4-10
4.4.2 Impulse response and transfer function of the NLoS component 4-10
4.4.3 Derivation of NLoS transfer function 4-13
4.5 Performance analysis of the V2V-VLC channel model 4-15
4.5.1 The received optical power and SNR 4-15

	4.5.2 Calculating the BER	4-16
	4.5.3 Time dispersion parameters for the channel	4-17
4.6	Simulation and results	4-18
4.7	Chapter summary	4-25
	References	4-25
5	**Adaptive optical OFDM for V2X communications**	**5-1**
5.1	The OFDM principle	5-1
5.2	Optical (unipolar) OFDM	5-4
5.3	Related work in adaptive modulation for VLC	5-5
5.4	Optical OFDM scheme of a V2V-VLC system	5-7
	5.4.1 DCO-OFDM modulation scheme	5-8
	5.4.2 ACO-OFDM modulation scheme	5-11
5.5	Performance analysis and bit-loading algorithm	5-12
	5.5.1 SNR estimation of subcarriers using singular value decomposition (SVD)	5-12
	5.5.2 Adaptive transmission and bit-loading scheme	5-13
5.6	Results and discussions	5-14
	5.6.1 Simulation environment	5-14
	5.6.2 Results and analysis	5-16
5.7	Chapter summary	5-19
	References	5-20
6	**Precoder and equalizer in 2 × 2 MIMO-VLC systems**	**6-1**
6.1	Precoding basics	6-1
	6.1.1 Zero forcing	6-1
	6.1.2 Minimum mean square error	6-2
	6.1.3 Maximum likelihood	6-2
6.2	Precoding and equalization for VLC	6-3
	6.2.1 Related work	6-5
	6.2.2 Dimming control technique and its performance in VLC systems	6-5
6.3	Precoder and equalizer in 2 × 2 MIMO V2V-VLC systems	6-7
	6.3.1 Investigating the transmitter	6-7
	6.3.2 Investigating the receiver	6-8
	6.3.3 Equalization matrix design	6-10
	6.3.4 Precoding matrix	6-11
6.4	Simulation and results	6-12

6.5	Chapter summary	6-13
	References	6-14
7	**Sunlight effects and denoising schemes**	**7-1**
7.1	Solar irradiance	7-1
7.2	SNR of V2V-VLC system	7-2
7.3	Related work in denoising schemes for VLC	7-5
7.4	Noise calculations	7-6
	7.4.1 Other noise sources	7-7
	7.4.2 Performance metrics	7-7
7.5	Differential receiver as a denoising scheme to improve the performance of V2V-VLC Systems	7-7
	7.5.1 Differential filtering scheme for 2 × 2 MIMO-V2V-VLC	7-8
	7.5.2 Signal filtration by differential receiver	7-9
	7.5.3 Simulation results	7-11
7.6	Denoising of V2V-VLC systems using machine learning	7-13
	7.6.1 System model	7-13
	7.6.2 Machine learning and adaptive filtering	7-14
	7.6.3 Why *k*NN algorithm	7-15
	7.6.4 Problem classification	7-15
	7.6.5 Adaptive filtering	7-15
	7.6.6 *k*-nearest neighbour algorithm and distance weighted *k*NN rule	7-16
	7.6.7 Simulation results	7-18
	7.6.8 Execution time	7-21
7.7	Chapter summary	7-21
	References	7-21
8	**Hybrid channel-based foglet-assisted smart asset reporting**	**8-1**
8.1	Introduction	8-1
8.2	System model	8-3
	8.2.1 VLC channel between the vehicle and the foglets	8-4
	8.2.2 Radio frequency link	8-5
8.3	Performance analysis	8-6
8.4	Results and discussion	8-7
8.5	Chapter summary	8-9
	References	8-9

Preface

The increasing scarcity of the radio frequency (RF) spectrum has spurred interest in light-based wireless communications for several years[1]. Optical wireless communication (OWC) encompasses wireless communication that employs light waves, both visible and invisible, to convey information[2]. Visible light communication (VLC), a subset of OWC, specifically utilizes the visible light spectrum [430–770 THz] for data transmission and can concurrently provide illumination.

The swift advancement of high-power LEDs within the visible spectrum has significantly accelerated interest in VLC. Note OWC, and particularly VLC, is emerging as a compelling alternative to RF communications for numerous reasons. Here are some of the key benefits:

- **High bandwidth:** OWC systems can support very high data rates due to the high frequency of optical signals, allowing for the transmission of large amounts of data.
- **High security:** The directionality and narrow beam of optical signals make it difficult for eavesdroppers to intercept the transmission, providing enhanced security.
- **Interference immunity:** Optical signals are not affected by electromagnetic interference (EMI) from RF sources, making them ideal for environments with high RF noise.
- **Unlicensed spectrum:** OWC systems operate in the optical spectrum, which is unlicensed and free to use, unlike RF bands that require regulatory approval and licensing.
- **High spatial reuse:** The highly directional nature of optical signals allows for high spatial reuse, enabling multiple links to coexist in close proximity without interfering with each other.
- **Low power consumption:** Optical communication systems, especially those using LEDs and photodetectors, can be more energy-efficient compared to traditional RF systems.
- **Ease of deployment:** OWC systems can be deployed quickly and easily without the need for extensive infrastructure, making them suitable for temporary or mobile setups.
- **No spectrum shortage:** The optical spectrum is vast and not crowded, unlike the RF spectrum, which is increasingly congested.
- **Safe for human exposure:** Optical signals, particularly those in the visible light range, are safe for human exposure and do not pose health risks associated with RF radiation[3].

[1] This is different from fiber optic communications.

[2] Free-space optical (FSO) communication is a prevalent method that generally denotes point-to-point communications using light beams.

[3] IEC825 is a standard that regulates the safety of OWC systems.

- **Versatility:** OWC can be used for a variety of applications, including indoor lighting-based communications (Li-Fi), underwater communication, and space communication.

These advantages make OWC an attractive option for a range of communication scenarios, complementing or even replacing traditional RF-based wireless systems in certain applications.

Consequently, the use of OWC, particularly VLC in automotive applications, as an alternative or supplement to traditional RF-based communications, has garnered significant interest.

This book investigates the suitability of VLC, especially in vehicular environments, in the realms of autonomous vehicle (AVs) and intelligent transportation system (ITS) as an emerging technology for increasing road safety and for improving traffic efficiency. Mainly, the book covers VLC-based vehicle-to-vehicle (V2V) considering both line-of-sight (LoS) and non-line-of- sight (NLoS) signal reception scenarios. This book presents challenges of communication between light emitting diode (LED)- based VLC emitters, that can typically be vehicle headlights or taillights and photodiode (PD)-based VLC receivers. A 2×2 multiple-input–multiple-output (MIMO) communication link is assumed, as appropriate, between a pair of transmitters and receivers.

The book includes seven chapters that cover the important aspects of V2V-VLC scientific theory and technology. Following this preface, chapter 1 presents a general introduction about VLC. chapter 2 presents a (2×2) MIMO channel for V2V-VLC links by considering both LoS and NLoS propagation scenarios. We analyze the effects of intermediate reflectors considering infinitesimal Lambertian reflecting elements. Then, the VLC channel impulse response (CIR) is obtained by combining received signals from these numerous reflectors in each reflecting surface. Multiple reflections are also considered. Afterwards, the obtained CIR is applied to investigate certain statistical properties of the channel such as the received power profile, mean access delay, RMS delay spread, coherence bandwidth and coherence time. This is done for vehicles moving in the same direction as well as in opposite directions.

Once we have a better understanding of the channel, certain advanced modulation schemes can be investigated. In chapter 3, optical orthogonal frequency division multiplexing (O-OFDM) with an adaptive modulation scheme is presented to mitigate inter-symbol interference (ISI), and also to improve data rate. Note, since the OWC environment is a *Unipolar* environment, bipolar OFDM signaling cannot be directly applied in the VLC environment. The negative amplitude signal has to be removed somehow. For this, we consider direct current optical (DCO)-OFDM and asymmetrically clipped optical (ACO)-OFDM schemes and compare their performances. DCO-OFDM is a form of O-OFDM with an appropriate DC bias added to lift-up the negative portion. In the ACO-OFDM technique, the modulated data is assigned to the odd indexed subcarriers, while the even indexed subcarriers are set to zero. This also removes the negative portion. Numerical simulations show data rates up to 50 Mbps with low bit error rate (BER) can be achieved with these algorithms under good conditions.

Note, one major challenge is that the outdoor vehicular VLC channels change rapidly and have frequency selective nature. Hence, in chapter 4, we investigate a precoding and equalizing scheme for the V2V-VLC system considering flickering/dimming control. We present the precoding matrix, as well as the equalization matrix for zero forcing (ZF), maximum likelihood (MLH), and minimum mean-square error (MMSE) receiving schemes. The MMSE scheme shows a noticeable improvement over other schemes.

Directional nature of light wave propagation creates shadows that could impair communications. Chapter 5 deals with the shadowing effect on the V2V-VLC system. We propose a method to take advantage of the optical diffraction phenomenon to overcome shadowing effect by employing a receiver with a wide field of view (FOV). The shadowing effect for visible light is modeled by a bimodal distribution and the probability of error for different types of shadowing is derived.

A major problem of outdoor VLC communication systems is the sunlight. The solar irradiance drastically effects the V2V-VLC system during daytime. Since this is overlapping in-band noise, it cannot be easily filtered out using optical/electrical bandpass filters. Different denoising schemes are investigated chapter 6. The performance of the system is investigated in terms of signal-to-noise ratio (SNR), BER, and data rate under three different scenarios. We proposed filtering the solar irradiance by using a differential receiver that works with two LEDs each with slightly different wavelengths. The performance of this scheme is compared with just an optical filter in front of the receiver. Also, we propose denoising the solar irradiance in a VLC link using a machine learning (ML) approach. We propose a *k*-nearest neighbor (*k*NN) algorithm-ML-based adaptive filter to combat the effect of solar irradiance. A supervised algorithm was used in the analysis. Both denoising schemes are discussed in detail.

The integration of artificial intelligence (AI) and ML in vehicular networks, commonly referred to as the internet of vehicles (IoV), brings transformative advantages that significantly enhance autonomous vehicles and intelligent transportation systems (ITS). One of the primary benefits is improved decision-making capabilities. AI and ML algorithms can analyze vast amounts of data from various sources in real time, such as traffic conditions, weather, and the behavior of other vehicles. This enables autonomous vehicles to make informed decisions rapidly, leading to safer and more efficient navigation. For instance, AI can predict potential hazards and suggest alternative routes to avoid accidents or traffic jams, thereby improving overall road safety and reducing travel time.

Another key advantage is the enhancement of vehicle-to-everything (V2X) communication. AI and ML facilitate seamless communication between vehicles, infrastructure, and other road users. This interconnectedness allows for the exchange of critical information, such as sudden braking by a vehicle ahead or changes in traffic signal status, enabling vehicles to respond proactively to dynamic driving conditions. This leads to a more synchronized traffic flow and reduces the likelihood of collisions. Additionally, AI-driven predictive maintenance can analyze data from vehicle sensors to anticipate and address mechanical issues before they result in breakdowns, thereby enhancing vehicle reliability and lifespan.

Furthermore, AI and ML contribute to the optimization of traffic management systems. Intelligent algorithms can process data from various traffic sensors and cameras to manage traffic lights and signage dynamically. This results in smoother traffic flow and reduced congestion, particularly in urban areas. Moreover, AI can identify patterns and trends in traffic data, helping city planners to design better infrastructure and implement effective traffic control measures. In the context of ride-sharing and public transportation, AI can optimize routes and schedules based on real-time demand, improving service efficiency and passenger satisfaction.

In summary, the incorporation of AI and ML into vehicular networks and ITS brings significant advancements in safety, efficiency, and reliability. By enabling real-time data analysis and decision-making, enhancing V2X communication, and optimizing traffic management, these technologies pave the way for smarter, more autonomous, and highly efficient transportation systems. As a result, they play a crucial role in the evolution of modern transportation, addressing current challenges and setting the foundation for future innovations. The newly added chapter 7 covers AI and ML in VLC systems.

Acknowledgements

Authors would like to express their heartfelt gratitude to all those who supported and contributed to the completion of this book. Without their guidance, encouragement, and assistance, this project would not have been possible.

First and foremost, we would like to thank Jonathan Diller, Illanko Kandasamy, Christopher Mekhiel, Noha Hassan and Fatima Hussain for allowing us to use portions of the co-authored material in this book. Your contributions in exploring theoretical and practical applications of VLC in vehicular environments were instrumental in shaping the discussions presented in this book.

We are also grateful to our colleagues and peers for their constructive suggestions and critique. The intellectual exchanges and diverse perspectives have significantly enhanced the quality of this work.

Moreover, we would like to thank our families and friends for their patience and support during the lengthy process of researching and writing this book. Your belief in our vision kept us motivated even during challenging times.

Lastly, we acknowledge the various academic institutions and organizations that provided resources, research materials, and a platform for our learning. Their commitment to promoting knowledge and innovation in the fields of communication and technology continues to inspire future generations.

Author biographies

Xavier Fernando

Xavier Fernando (https://www.ecb.torontomu.ca/~fernando) is a Professor and Director of Intelligent Computing and Communications Lab at Toronto Metropolitan University (TMU), Canada. He was an IEEE Distinguished Lecturer and delivered over 70 lectures worldwide. He has (co-)authored over 200 research articles, 3 books, several book chapters and holds three patents. He was a member of TMU Board of Governors 2010–11. He is a program evaluator for ABET and has visited American Universities for accreditation. He was a finalist for the Top 25 Immigrant Award of Canada in 2012. He is very active in IEEE (Institute of Electrical and Electronic Engineers). He was the Chair of IEEE Humanitarian Technology Initiatives Committee, Canada. He was the General Chair of *IEEE International Humanitarian Technology Conference*, 2017. He was the General Chair for *IEEE Canadian Conference on Electrical and Computer Engineering (CCECE)* 2014 and 2026. He was the Chair of IEEE Toronto section and Member of IEEE Canada Board during 2012–13 and IEEE Central Canada Chair during 2015–16. He and his students have won 32 awards and prizes so far including; the First prize at the *Humanitarian Initiatives Workshop of CCECE 2014*; Second Prize at the *2014 IHTC Conference in Montreal*; Best Paper Award at the *International Conference of Smart Grid Engineering (SEGE 2014)*, UOIT, Canada, 2014; IEEE Microwave Theory and Techniques Society Prize in 2010, *Sarnoff Symposium* prize in 2009, Opto-Canada best poster prize in 2003 and CCECE best paper prize in 2001. Xavier Fernando earned his PhD from the University of Calgary, Alberta in 2001 and BSc (Eng) from the University of Peradeniya, Sri Lanka in 1993.

Hasan Farahneh

Hasan Farahneh (https://www.researchgate.net/profile/Hasan-Farahneh) has been on the teaching staff at the University of Jordan since 2004. He received a BS from Al-Yarmouk University, Irbid, Jordan in 1985, an MSc from the University of Jordan, Amman, Jordan in 2004, and a PhD from Toronto Metrpolitan (formerly Ryerson) University Toronto-Canada in 2018. In September 2018, he joined the Electrical Engineering Department at the University of Jordan as a teaching staff member. He is an IEEE (Institute of Electrical and Electronic Engineers) member and a member of Jordan Engineering Association. His research interest includes wireless and visible light communication. His current research interest is focused on vehicular communication using visible light spectrum and artificial intelligence in optical communications.

List of acronyms

ACO-OFDM	Asymmetrically Clipped Optical-Orthogonal Frequency Division Multiplexing
A/D	Analog-to-Digital
ADAS	Advanced Driver Assistance System
AI	Artificial Intelligence
AODV	Ad Hoc On-Demand Distance Vector
APD	Avalanche Photodiode
AV	Autonomous Vehicle
AWGN	Additive White Gaussian Noise
BER	Bit Error Rate
BPF	Band Pass Filter
BPSK	Binary Phase Shift Keying
BRDF	Bidirectional Reflectance Distribution Function
CDMA	Code-Division Multiple Access
CDSF	Connected Dominating Set-based Forwarding
CIR	Channel Impulse Response
CP	Cyclic Prefix
CMOS	Complementary Metal-Oxide Semiconductor
CSI	Channel State Information
C-V2X	Cellular Vehicular-to-Everything
D/A	Digital-to-Analog
DCO-OFDM	Direct Current Optical—Orthogonal Frequency Division Multiplexing
DFT	Discrete Fourier Transform
DMT	Discrete Multi-Tone Modulation
DPC	Dirty Paper Coding
DSRC	Dedicated Short-Range Communications
DSSS	Direct Sequence Spread Spectrum
EMI	Electromagnetic Interference
FFT	Fast Fourier Transform
FOV	Field of View
FDM	Frequency Division Multiplexing
FSO	Free-Space Optical
IDFT	Inverse Discrete Fourier Transform
IF	Intermediate Frequency
IFFT	Inverse Fast Fourier Transform
IM/DD	Intensity Modulation/Direct Detection
IoT	Internet of Things
IoV	Internet of Vehicles
IR	Infrared
ISI	Inter Symbol Interference
ITS	Intelligent Transportation Systems
IVC	Inter-Vehicular Communication
I2V	Infrastructure-to-Vehicle
kNN	k-Nearest Neighbor
LED	Light Emitting Diode
LEO	Low Earth Orbit.
LIDAR	Light Detection and Ranging Sensor

Li-Fi	Light Fidelity
LoS	Line-of-Sight
LPF	Low Pass Filter
LWR	Lighthill–Whitham–Richards
MIMO	Multiple-Input–Multiple-Output
MLH	Maximum Likelihood
ML	Machine Learning
MMSE	Minimum Mean-Square Error
MSE	Mean Squared Error
NHTSA	National Highway Traffic Safety Administration
NLoS	Non-Line-of-Sight
NLP	Natural Language Processing
OBU	On Board Unit
OCI	Optical Communication Image Sensor
OFDM	Orthogonal Frequency Division Multiplexing
O-OFDM	Optical-Orthogonal Frequency Division Multiplexing
OOK	ON–OFF-Keying
OW	Optical Wireless
OWC	Optical Wireless Communication
PAM	Pulse Amplitude Modulation
PD	Photodiode
PDF	Probability Density Function
PDs	Photodiodes
PIN	Positive Intrinsic Negative Photodiodes
PPM	Pulse Position Modulation
PSD	Power Spectral Density
PWM	Pulse Width Modulation
P/S	Parallel-to-Serial
QAM	Quadrature Amplitude Modulation
QoS	Quality of Service
RF	Radio Frequency
RMS	Root Mean Squared
ROI	Region of Interest
RSSI	Received Signal Strength Indication
RSU	Road Side Units
RZ	Return to Zero
SISO	Single-Input–Single-Output
SNR	Signal-to-Noise-Ratio
S/P	Serial-to-Parallel
SRC	Safety Related Critical
SSL	Self-supervised learning
STBC	Space-Time Block Coding
SVD	Singular Value Decomposition
TDE	Time Domain Equalization
THP	Tomlinson–Harashima Precoding
UAV	Unmanned Aerial Vehicles
UMBP	Urban Multi-hop Broadcast Protocol
VLC	Visible Light Communication
VANET	Vehicular Ad-hoc Network
V2E	Vehicle-to-Environment

V2I	Vehicle-to-Infrastructure
V2N	Vehicle-to-Network
V2P	Vehicle-to-Pedestrian
V2V	Vehicle-to-Vehicle
V2X	Vehicle-to-Everything
V-IoV	Visual IoV
WAVE	Wireless Access for Vehicular Environments
WDM	Wave Division Multiplexing
WSM	WAVE Short Message
ZF	Zero Forcing

IOP Publishing

Visible Light Communications (Second Edition)
Vehicular applications
Xavier Fernando and Hasan Farahneh

Chapter 1

Introduction

This chapter discusses optical wireless communication (OWC) and its application in vehicular environments, particularly focusing on visible light communication (VLC). OWC encompasses technologies that transmit data via light waves, offering a viable alternative to traditional radio frequency (RF) communications, especially in scenarios where RF is congested or subject to interference. VLC utilizes the existing LED lighting in vehicles and infrastructure, enabling high-speed, short-range data exchange among vehicles (V2V), between vehicles and infrastructure (V2I), and with networks (V2N) or pedestrians (V2P). It highlights benefits such as improved collision avoidance, traffic management, and enhanced positioning precision. This chapter also acknowledges challenges like interference from sunlight, quick channel changes, and the necessity for line-of-sight communication. Overall, it underscores the potential of VLC in enhancing intelligent transportation systems (ITS) and supporting the development of autonomous vehicles through reliable, real-time communication. The exploration of research directions and practical applications of VLC aims to address these challenges while maximizing the benefits of this emerging technology.

1.1 Optical wireless communication

OWC, encompassing infrared (IR) and VLC, refers to the wireless transmission of data through light waves without the need for optical fibers. VLC, in particular, is a rapidly advancing technology that utilizes the visible light spectrum. This technology holds significant promise as a wireless alternative, providing increased bandwidth, energy efficiency, security, and data density, without being affected by or causing electromagnetic interference (EMI)[1].

[1] Except for the characteristic of falling within the visible spectrum, most attributes of VLC are identical to those of OWC. Therefore, in this text, the terms OWC and VLC are used to describe these general characteristics as they fit.

doi:10.1088/978-0-7503-6049-4ch1

It is important to recognize that VLC can fulfill a dual role, providing both illumination and data transmission simultaneously. The growing interest in VLC is due to the extensive use of solid-state lighting, such as LEDs, in buildings and vehicles. LEDs are capable of rapid switching on and off, or modulation with a digital or analog signal. The circuits required to drive and modulate LEDs are significantly simpler and more cost-effective than those required for other lighting types.

Consequently, simple and economical intensity modulation/direct detection (IM/DD) techniques are particularly appealing in OWC. In intensity modulation, modulation is limited to signal intensity without transmitting phase information. Direct detection is confined to the optical power received. As a result, communication signals in IM/DD systems are real-valued and *unipolar* (non-negative). Therefore, advanced modulation and signaling schemes from RF wireless communications, such as orthogonal frequency division multiplexing (OFDM) and code-division multiple access (CDMA), cannot be directly applied in unipolar IM/DD contexts. They require specific modifications to eliminate the negative part of the waveform. It should also be noted that multipath fading does not occur in an IM/DD environment since the power is always additive, although multipath dispersion and inter-symbol interference (ISI) may still be present.

While RF-based communication systems are becoming increasingly saturated with a large number of internet of vehicle (IoV) nodes and users due to limited spectrum availability, OWC systems emerge as a compelling alternative. OWC not only offers a vast bandwidth that is license-free and thus available at no cost, but it also inherently possesses directional propagation properties, typically encompassing a short range. Although some may view these characteristics as limitations, they significantly reduce the likelihood of interference with other communication links and diminish the risk of eavesdropping within OWC environments. Furthermore, OWC systems are not susceptible to and do not contribute to electromagnetic interference, which is common in RF communication systems. These features render OWC systems an optimal choice for environments sensitive to electromagnetic disturbances, such as aircraft, hospitals, and areas near power lines. Additionally, the propagation properties of visible light in OWC systems are highly advantageous for underwater communication applications.

OWC offers significant advantages for vehicular communications for several reasons:

- Most vehicles come equipped with LED head and taillights, which can be readily converted into VLC transmitters with minimal modifications.
- Image sensors, already installed in most vehicles, can also be adapted to serve as VLC receivers with the right modifications.
- The communication between nearby vehicles, particularly those directly ahead or behind, can be easily established due to the short distances and relatively stable channel conditions. These short-range VLC links do not interfere with vehicles that are further away.
- Most traffic lights, now equipped with LEDs, have the capability to communicate with vehicles within their line of sight. This technology

introduces an additional layer for communication and the potential for short message transmission [1].

- During traffic congestion when vehicles move slowly, VLC can be particularly beneficial as RF communication networks, like IEEE 802.11p, may become overloaded with excessive connection requests. In such scenarios, hybrid RF/VLC approaches can ease the situation, enabling transmission of emergency messages to nearby vehicles [2].
- Improved positioning: The directional nature of the VLC beams helps in determining the sender and receiver positions more precisely than in the case of non-directional RF. This helps in the routing and emergency response.
- Supplements RF gaps: The VLC can relay the messages between the vehicles/ infrastructure in situations. In such case: the RF signals may be obstructed, such as tunnels and underground parking structures.

Furthermore, vehicular VLC can be categorized into V2V, vehicle-to-infrastructure (V2I), vehicle-to-network (V2N) or vehicle-to-pedestrian (V2P) communications. OWC can also be used for in-vehicle communications providing passengers with high-speed internet access, video streaming, and other services without the need for RF-based Wi-Fi.

V2V communication: VLC can enable high-speed communication between vehicles using their headlights, taillights, and other lighting systems. This can facilitate:

- Collision avoidance: real-time exchange of speed, direction, and location data to prevent accidents.
- Platoon formation: coordinated driving of multiple vehicles close together to improve fuel efficiency and reduce traffic congestion.
- Overtaking assistance: sharing information about the road ahead for safer overtaking maneuvers.

V2I communication: VLC can be used to communicate with traffic signals, streetlights, and other infrastructure. This can enhance:

- Traffic management: vehicles can receive real-time updates on traffic conditions, road hazards, and optimal routes.
- Signal priority: emergency vehicles can communicate with traffic lights to get green signals along their route.
- Parking assistance: OWC-enabled parking spaces can guide vehicles to available spots and assist with automated parking.

V2N communication: a communication paradigm where vehicles connect to cellular networks to communicate with network entities such as cloud servers, traffic management systems, and other infrastructure.

V2P communication: VLC can enhance safety for pedestrians through:

- Crosswalks and signals: OWC-enabled crosswalk signals can communicate with vehicles to alert drivers when pedestrians are crossing.
- Personal devices: pedestrians with OWC-enabled devices can receive warnings from nearby vehicles about potential dangers.

However, interference from intense sunlight, swiftly changing channel conditions, extended distances between transmitters and receivers, and the requirement for line of sight channels present significant challenges for vehicular VLC. This book delves into various such scenarios, examines the associated challenges, and explores in-depth research solutions.

1.1.1 Cellular V2X for intelligent transportation systems

For autonomous driving and ITS, real-time vehicular communication is crucial. This is particularly true in public traffic, where multiple vehicles with varying velocities and positions interact. The high volume of traffic and limited lane resources necessitate frequent interactions between neighboring vehicles, leading to interdependent vehicle motion. Reliable, high-speed wireless communication networks with very low latency are required to communicate immediate navigational intentions and understand those of other vehicles.

Hence, vehicle-to-everything (V2X) communications, where 'X' represents vehicles, infrastructure, networks, cellular base stations, road side units (RSUs) like traffic lights, or pedestrians, are vital for both ITS and autonomous vehicles. Recognizing its significance, 5G wireless network standards have introduced specific features tailored to support these requirements such as *ultra-reliable* and *low-latency* modes.

The foundational V2X standard derives from IEEE 802.11p, a Wi-Fi variant and part of IEEE's WAVE, also known as DSRC, operating in the unlicensed 5.9 GHz frequency band. This standard encompasses sensors like cameras, radar, and light detection and ranging (LiDAR), and supports V2V and V2I applications such as collision warnings, speed limit alerts, and electronic parking and toll payments. The 802.11p's functional traits include a short-range (less than 1 km), low latency (approximately 2 ms), and high reliability. The US Department of Transportation states it is effective in high-speed vehicular conditions and resistant to extreme weather conditions like rain, fog, and snow. In essence, 802.11p enhances a vehicle's capacity to perceive its surroundings, even in challenging weather conditions.

Uninterrupted wireless connectivity is a key enabler of seamless information flow to and from densely deployed vehicles, infrastructure, cloud, and IoV applications. Furthermore, vehicles need to communicate with everything around them reliably. without failures. In mission-critical applications such as connected cars, the availability and reliability requirements to prevent catastrophic communication link failure need a trade-off with highly flexible, configurable, scalable and reverse-compatible data formats (figure 1.1).

The transmission of information between vehicles and any entity that may have an impact on the vehicle, such as other vehicles (V2V), infrastructure (V2I), pedestrians (V2P), wireless networks (V2N) and the environment (V2E), is referred to as V2X communications. V2X communication is essential even if the vehicle has its own sensor system such as advanced driver assistance system (ADAS). ADAS have prioritized vision systems for detecting objects on a vehicle's path. While these systems are highly advanced and capable of recognizing various obstacles, they

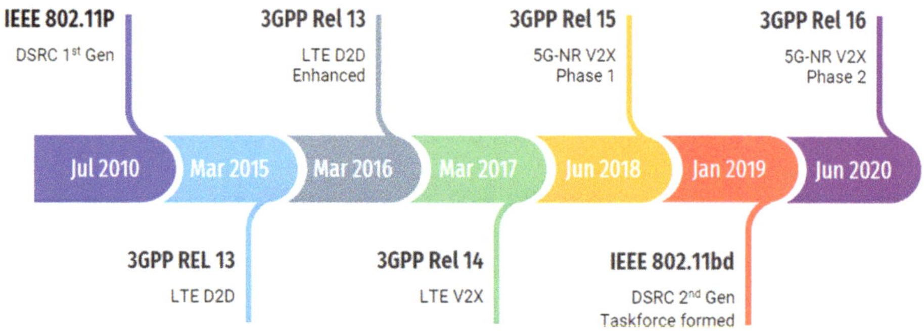

Figure 1.1. Evolution of V2X communication standards. From 3GPP TSG SA#87e, e-meeting document SP200222[2].

cannot see around corners. To address this, fixed hazards such as railroad crossings or road construction could be equipped with V2X transmitters, allowing vehicles to detect them before they come into view, even in conditions like fog or behind curves. V2X systems not only enhance safety but also comfort and convenience. For example, parking lot sensors utilizing V2X technology can detect vacant spots and direct vehicles to them [3].

The 5G wireless technologies such as cloud radio access networks (C-RAN), MIMO, and coordinated multipoint (CoMP), common public radio interface (CPRI), and the next-generation front-haul interface (NGFI) are expected to act as a platform for novel and promising applications of V2X in 5G and 6G IoV networks.

V2X can also help with economy; when an area is crowded, the satellite navigation system can use this information to allow—or force—the vehicle to avoid the crowded area. Similar to this, air quality sensors positioned in densely pedestrianized areas have the ability to detect deteriorating air quality and send out signals to internal combustion engine cars to reroute. V2X systems will develop over time, with the majority of their early iterations serving as warning signals for drivers. The technology will gradually take over control of the vehicle's movement as it becomes increasingly integrated with the vehicle's systems (figure 1.2).

The major advantage of cellular V2X (C-V2X) is that it has two operational modes:

- **Low-latency PC5 or sidelink mode:** This C-V2X direct communication mode is designed for active safety messages, such as immediate road hazard warnings. This is akin to IEEE 802.11p system. This mode enables direct V2V, V2P and V2I communications. It operates independently of cellular networks, using the 5.9 GHz band.
- **Network/Uu interface mode:** This mode leverages existing cellular networks (4G LTE, and 5G) to facilitate mainly V2N communications, providing access to a wide range of information and services, such as cloud-based applications, traffic management systems, and infotainment. The Uu

[2] https://www.3gpp.org/dynareport?code=spexmtg-SP-87-e.htm

Figure 1.2. Overview of V2X communications. © 2019 IEEE. Reprinted, with permission, from [4].

interface supports longer-range communication and high data rates, making it suitable for non-safety applications that require internet connectivity.

1.1.2 Visible light technology in V2X communications

VLC offers reliable connection complementing radio links. In particular, OWC and VLC can play a vital role in communicating with flying or aerial objects such as unmanned aerial vehicles (UAV)s that enable aerial-terrestrial communication networks[3].

Emerging optical communication technologies such as 100+Gb/s, coherent and low-cost IM/DD allow high-throughput and low-latency V2X communication. The IM/DD transmission leads to smaller VLC transceiver form factors to meet increasing demands on network capacity, and surge in data traffic. AI-driven design, optimization, and network management for VLC MIMO communications requires efficient channel estimation and symbol detection, along with modulation and coding scheme design and resource allocation that leads to these intelligent environments measuring, analyzing, and controlling themselves in an automated manner. The VLC capability maximizes the IoV network capacity with more short-range connections with more access points.

The role of visible light in V2X communications can be appear in different aspects and stages of V2X such as [5]:

- **Data transmission:** Various emergency and non-emergency information pertinent road conditions, traffic intensity etc can be transmitted to other vehicles and RSUs via VLC.

[3] The International Telecommunication Union telecommunication standardization sector (ITU-T) is working on a set of official recommendations for aerial-terrestrial communication based Internet of Space Things [6].

- **Localization and positioning:** In areas like urban canyons or tunnels where GPS signals may be erratic or nonexistent, visible light can be used for accurate localization and positioning. Vehicles may precisely locate themselves by using visible light beacons that are placed in infrastructure or along highways.
- **Huge bandwidth:** Because visible light can adjust light intensity at high frequencies and has a vast accessible spectrum, it offers a potentially high bandwidth for communication. Large-scale data transfer between vehicles and infrastructure is made possible by this, including high-definition video and sensor data.
- **Low interference:** Distinct from typical RF communication technologies like Wi-Fi or cellular networks, visible light communication operates in a distinct spectrum. By reducing traffic and interference in the RF range, this can promote more dependable and effective communication between moving cars and their environment.
- **Security:** Since visible light signals are limited to a physical area and are less vulnerable to jamming or interception than radio frequency signals, they can offer a higher level of security than RF communication. For V2X communication applications that are safety-critical, such as collision avoidance systems, this can be extremely crucial.
- **Energy efficiency:** Because LED-based visible light communication systems are simple to incorporate into the current lighting infrastructure, they are by nature energy-efficient. This implies that visible light-based V2X communication can take advantage of the current infrastructure without dramatically consuming more energy.

Therefore, it is clear that VLC is anticipated to play a bigger and bigger part in enabling effective and dependable communication between vehicles and their surroundings as research and development in this area continues.

VLC techniques are particularly beneficial in certain scenarios, such as communication with closely trailing vehicles and RSUs. While LTE-based V2X and 802.11p are apt technologies for supporting cooperative awareness, sensing, and maneuvering, IEEE 802.11p's performance suffers in dense, high-traffic network environments. Moreover, the current functions and procedures of LTE-based systems are not designed to fulfill the stringent delay and reliability requirements necessary for advanced cooperative sensing and maneuvering.

Autonomous vehicles require strict adherence to delay and reliability standards that the present LTE-based cellular systems cannot provide, necessitating the consideration of next-generation cellular systems for these applications. Conversely, VLC is ideally suited for use cases like see-through and platooning, which involve short-range, line-of-sight communication between a leading and following vehicle.

1.1.2.1 Li-Fi (light fidelity)

Li-Fi is defined as an optical wireless broadband access technology that uses the visible and infrared light spectrum to provide bidirectional wireless communications.

It supports simultaneous lighting and communications in the VLC band and also point-to-point or point-to-multipoint topology in both forward and reverse directions, and provides multiuser access as well. Li-Fi can also offer mobility support for both intra-cell and inter-cell movement. There are two principal versions of handover:

1. Horizontal handover: The ability to maintain a data connection to the network as a user moves from one cell to another using the same communications technology (Li-Fi).
2. Vertical handover: The ability to maintain a data connection to the network as a user changes the principle radio access technology, i.e., moving from 4G/5G to Li-Fi or from Wi-Fi to Li-Fi.

The Li-Fi infrastructure consists of multiple LEDs that create a wireless network, providing a typical MIMO environment. Developers of Li-Fi systems foresee interoperability with Wi-Fi, 4G, or other RF systems, allowing users to connect 'opportunistically' and even simultaneously to both Li-Fi and other RF wireless networks.

Li-Fi can enable high-speed communication between vehicles using their headlights, taillights, and other lighting systems. Li-Fi can also be used to communicate with traffic signals, streetlights, and other infrastructure, and it can be used for in-vehicle communication supporting infotainment systems.

However, Li-Fi requires a clear line of sight between the transmitter and receiver, which can be a limitation in certain scenarios. Weather conditions like fog, rain, and snow can affect the transmission of light signals. Overall, the potential benefits of Li-Fi in vehicular applications make it a promising technology for future transportation systems, enhancing safety, efficiency, and connectivity.

1.1.2.2 IEEE 802.11bb™

In May 2018, IEEE announced the formation of the IEEE 802.11 Light Communications Task Group to develop a global standard for VLC in wireless local area networking, specifically to extend Wi-Fi (IEEE 802.11™) protocols to operate using visible light. The IEEE 802.11bb™ task group is to define one medium access control (MAC) and several physical layer (PHY) specifications for light-based wireless connectivity for fixed, portable, and moving stations within a local area network. The new standard will closely align with Li-Fi.

A key specification of IEEE 802.11bb™ is: uplink operation is in 380 nm band, while downlink operation is in the 380–5000 nm band. Also, all modes of operation of IEEE 802.11bb™ are supposed to achieve minimum single-link throughput of 10 Mb s^{-1}. Meanwhile, at least one mode of operation will achieve single-link throughput of at least 5 Gb s^{-1}, as measured at the MAC data service access point (SAP). Interoperability among solid-state light sources with different modulation bandwidths will be supported.

In summary, the IEEE 802.11bb™ s standard represents a significant advancement in wireless communication technology, leveraging the advantages of visible light for secure, high-speed data transmission. The future of IEEE 802.11bb™ is promising, with potential applications in various industries and continued technological advancements driving its adoption and integration into modern communication systems.

1.2 VLC for intelligent transportation

The number of vehicles using transportation infrastructure increases every year, making it essential to continuously improve the safety and efficiency of the transportation system. Despite advancements in vehicle safety, over 1.3 million people die annually, and 20 to 50 million are injured due to car accidents [7]. Reducing road accidents and their associated victims remains a critical concern.

Beyond car safety systems, there is a strong need for vehicle awareness, achieved through vehicles sharing information to enhance safety and create an efficient road accident prevention system. This requires high cooperation among vehicles and between vehicles and transportation infrastructure. ITS is a promising solution to road accidents. ITS involves intelligent vehicles and infrastructure working together to improve the safety and efficiency of the transportation system [8]. These systems use advanced wired and wireless communication technologies for data gathering and distribution. By enabling wireless communications among vehicles (V2V) and between vehicles and infrastructure (V2I), road traffic safety and efficiency can be significantly improved. Figure 1.3 shows the application scenario for vehicular VLC.

Since 1995, academic and industrial societies have been considering IVC systems. Recently, VLC has been proposed as a potential solution to enable IVC. The main advantages of using VLC in automotive applications include low complexity, reduced implementation cost, and its ubiquitous nature. These characteristics can facilitate rapid and widespread market penetration, making VLC a strong contender.

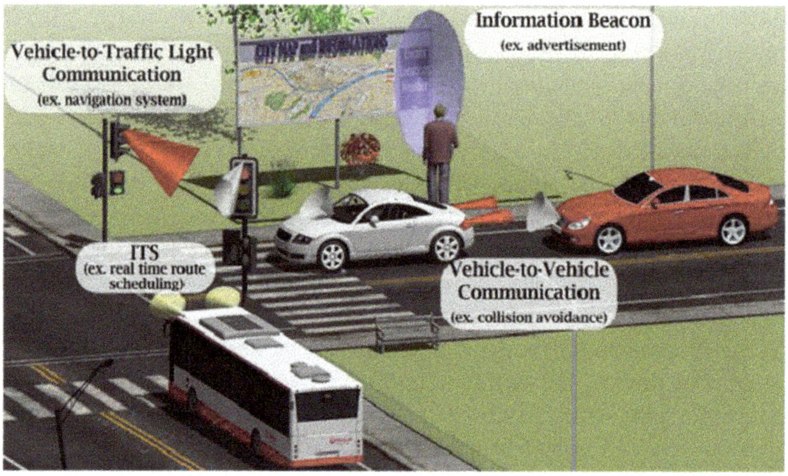

Figure 1.3. Application scenario for vehicular VLC. Copyright © 2011, IEEE. Reprinted, with permission, from [9].

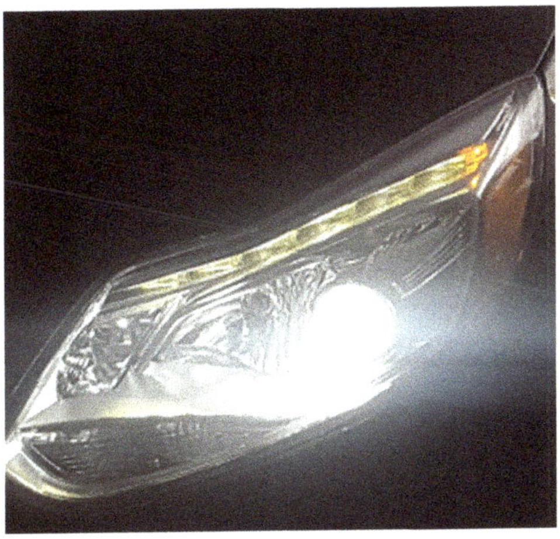

Figure 1.4. Integration of LEDs lighting systems in series vehicles.

Due to the features of LEDs, car manufacturers are considering replacing traditional halogen lamps with LED/laser lighting systems. LEDs are highly reliable, energy-efficient, and have a much longer lifespan compared to classical light sources. Nowadays, as illustrated in figure 1.4, LED-based vehicle lighting systems are common.

Moreover, LED-based traffic lights are becoming more popular and are being used on a larger scale. Typically, a large number (100–200) of LEDs are used in traffic lights, providing both signaling functions and the possibility for data transmission [1]. Additionally, most street lighting will soon be LED-based, allowing road illumination to support communication for infrastructure-to-vehicle (I2V)-VLC.

From the above, it is clear that LED-based lighting will be integrated into the transportation system, both in vehicles and infrastructure. The widespread use of LED lighting, combined with VLC technology, will enable ITS to gather data from a large area and distribute high-quality communications. These additional functions will not affect the primary goal of signaling or lighting. This demonstrates that VLC can meet the requirements of vehicular networks in real-world conditions.

1.2.1 VLC in the ITS/state of the art

Many researchers have investigated the use of VLC in automotive communication scenarios to enhance traffic safety. The results obtained so far prove that VLC can be used for road safety applications, but for widespread implementation the performance of such systems still needs more work and improvement. In the following paragraphs, we present a brief review of the work of some of the most representative research groups in the area of VLC automotive usage.

The Smart Lighting Engineering Research Center at Boston University is involved in using VLC for IVC. The research team has developed a prototype for vehicular networking based on optical transceivers. This system uses short-range directional optical transceivers to share vehicle state data. They also conducted a comparative analysis between omnidirectional 802.11 RF communications and directional VLC for vehicular communications [10]. The results show that in high traffic density, VLC offers better performance in terms of packet delivery ratio, throughput, and average packet delay, although it has a shorter communication range [11].

Professor Knightly's group from Rice University (USA) is another research team focused on vehicular communication using VLC. They conducted a detailed analysis of vehicular VLC-based networks and developed a research platform with high noise robustness. Their experiments in various conditions and scenarios demonstrated that VLC can provide robust vehicular communication in real traffic situations. The results showed that the V2V-VLC system is resilient to sunlight noise, except when there is a direct line of sight with the Sun. Additionally, their findings indicate that in dense traffic conditions, VLC meets the latency and reachability requirements necessary for vehicle safety applications [12].

G Pang from the University of Hong Kong and his research group are pioneers in VLC systems for traffic safety. They believe LEDs have the potential to replace traditional light sources and be integrated into traffic lights, signaling devices, and display boards due to their fast switching ability. This group proposed replacing all traffic lights and signaling devices with LEDs to reduce power consumption and increase traffic safety. They also highlighted the long life expectancy of LEDs, even in harsh conditions, as crucial since a burned-out traffic light can be a major risk factor. Additionally, the communication capabilities of LEDs can further enhance transportation safety. The authors demonstrated the dual use of LEDs in ITS: signaling and communication. To support their arguments, the research group presented a prototype traffic light that, besides traditional signaling, also transmits audio information through visible light [13]. Moreover, they proposed an intelligent traffic light system for broadcasting vehicle location and navigation information with audio support for drivers. The same group presented a different approach for traffic light-to-vehicle communication [14].

Researchers from Nagoya University in Japan are among the pioneers in the ITS field. They have a notable background in developing ITS-VLC systems. In 2005, they proposed a parallel VLC system designed to broadcast traffic safety information from an LED-based traffic light to a high-speed camera-based receiver [15]. In 2007, they introduced a novel VLC receiver prototype aimed at addressing key challenges in using VLC for ITS, such as the need for long-distance, high-speed transmission under dynamic conditions [16]. Experimental results showed a communication speed of 2 Mb s^{-1} with a BER below 10^{-6} for distances up to 40 meters. Over the years, the performance of high-speed camera-based receivers improved with advancements in coding schemes and camera technology. The communication distance increased to 120 meters, albeit with a higher BER of 10^{-2} [17].

The collaboration between Shizuoka University and the Applied Optics Laboratory at Toyota has led to the development of a high-performance VLC system. They have developed a high-sensitivity CMOS image sensor capable of achieving 1000 frames per second. Integrating this image sensor into a VLC receiver, their V2V prototype achieved data rates of 10 Mb s^{-1} for distances up to 20 meters. The communication range can be extended to 50 m by reducing the data rate to 32 kb s^{-1}, or even up to 100 m for data rates of 2 kb s^{-1} [18].

One of the leading groups researching VLC usage for automotive applications is at the University of Aveiro in Portugal. The group has proposed and analyzed in detail the use of LED-based traffic lights as RSUs as part of the ITS for I2V data broadcast. They have considered the use of PD-based receivers and proposed direct sequence spread spectrum (DSSS) inverse keying as a modulation technique [19]. Experimental results showed that this technique is suitable for outdoor VLC as it reduces noise from artificial light sources [20]. The system has been tested under outdoor conditions both during the day and at night. It can transmit data over distances of more than 40 m, with a BER between 10^{-6} at 10 m and 10^{-2} at 45 m, at a data rate of 20 kb s^{-1}.

1.3 AI-based V2X VLC communications

Recently, artificial intelligence (AI) and machine learning (ML) have been changing the world in an unprecedented manner. These technologies will immensely transform our daily lives via innovative means and new frontiers of communication such as connected vehicles, ITS, IoV and autonomous vehicles (AVs). Internet of Things (IoT) and IoV connectivity has led to a rapid increase in data communication in unforeseen ways such as reading the news on public commutes, utilizing digital maps for navigating destinations, and virtual reality-based navigation. For instance, wireless mobile data traffic has increased nearly 10-times whereas the peak data rate has increased by 55% in the last decade [21]. Standardization initiatives are aimed at integrating AI and ML as essential subsystems of networks beyond 5G and 6G in the realms of IoVs. 6G is expected to incorporate data rates of 100 Gb s^{-1} with 1 ms latency while also leveraging the existing 4G, and 5G networks. Note, low latency is critical for safe vehicular communications. The multi-band communication over radio frequencies (2–30 GHz), millimeter wave (30–300 GHz), sub Terahertz band (up to 1 THz), and optical frequencies (with typical bands of 850 nm/350 THz, 1300 nm/230 THz, and 1550 nm/193 THz) will disrupt physical, data-link and MAC layer designs of wireless networks. Vehicular safety, autonomous navigation and road traffic management use cases will be provided through V2X communications in which VLC will play an essential role. These technologies are seen as extremely promising for enhancing network functions related to control, management, and user experience as well as for supporting applications related to road safety and even entertainment.

Over the past few decades, the growing demand for urban mobility and modern logistics has led to a steady increase in the number of vehicles. This has resulted in severe traffic congestion in cities of all sizes, affecting economic expansion as well. Traffic incidents have increased significantly in the last five years, with estimates suggesting that at least one person dies in a road accident every minute globally [22]. In addition to traffic jams and accidents, other issues such as wasted working hours in traffic congestion during rush hour, difficulty in finding parking spaces and environmental pollution from vehicle emissions, including CO_2, dust particles, noise, and other pollutants that negatively impact the environment and public health have become more prevalent.

Recently, advancements in AI have enabled the development of smart vehicles, which aim to address these issues. Smart vehicles equipped with advanced sensors and connectivity features aim to improve road safety and reduce environmental impact [23]. These technologies enhance the ability of vehicles to navigate safely and efficiently, contributing to a safer and more sustainable transportation system. For instance, ADAS are found in many cars. These systems use different types of sensors to provide the driver or the controller in AVs with a variety of warnings. Although the inputs will be over-the-air communications, the V2X can greatly aid ADAS. V2X inputs will enable the ADAS system to respond to them in the same manner as it would to inputs from a nearby sensor.

Moreover, AVs are being deployed rapidly and can become a valuable option in transportation and the military. Self-driving AVs use various sensors and AI algorithms for their effective operation. They learn the environment leveraging ML techniques that fall under the umbrella of AI[4]. To have smart and effective planning for AVs development, intelligent software and tools are needed. These tools are utilized for mid-way planning, object detection, recognition, action, operational testing, and risk assessment stages. Thanks to various versions of intelligent software and tools such as AutoSim 200, OpenCV 4.5.5, and FuzzyJ 1.2.2, which are dedicated to cars, developers and researchers can rely on them to develop and evaluate their work [23].

VLC has garnered significant interest across vehicular industries due to its versatility and applicability. In the context of smart vehicles, VLC presents both opportunities and challenges. V2V infrastructure and network connection via VLC and its integration with ITS can enhance anti-collision systems, position-aware and location-based services, thereby improving safety, efficiency, and operational effectiveness in vehicular networks. This integration offers benefits such as traffic management, optimized routes and enhanced operational efficiency [24]. VLC can also seamlessly integrate with existing lighting infrastructure, providing an efficient solution for both illumination and communication [25].

VLC can contribute to ultra-reliable and low-latency services in 5G/6G network architectures as a complementing communication technology at various aspects of emerging IoV architectures. Modern vehicles are equipped with visual sensors, forming the visual IoV (V-IoV) that relies on computer vision and image processing techniques for sensing and processing visual data. These networks can even use autonomous aerial vehicles through intelligent recognition and visual information analysis. AI-driven IoV architectures for edge computing use analytics for multiple

[4] Despite the extensive research in the field of AVs, issues such as social, ethical, safety and security, regulatory, and technical challenges still need to be resolved and defined. Within the AVs framework, there is a high expectation to address all these issues to avoid disappointment in achieving excellent and safe AVs [26].

sensor data, multi-modal sensing and fusion to optimize various parameters to provide a smooth and smart next-generation transportation.

Another issue is the huge amount of data in IoVs. VLC links are increasingly being investigated to enable seamless transition from mm wave and sub Terahertz band links or to supplement them with near-zero latency to accommodate the exponential increase in data rates and the number of connected vehicles. Emerging advanced technologies such as holographic teleportation operate in a 3D space with data rate close to 5 Tb s^{-1} and may offer an end-to-end latency of less than 1 ms [21]. These kinds of derivatives of VLC will play an immense role in upcoming smart IoV.

1.4 The autonomous vehicle

Self-driving cars or driverless cars sometimes referred to as AVs. AVs are defined as vehicles that use a collection of cameras, LiDAR, radar and different types of sensors with appropriate software based algorithms to drive themselves from place to place without the need for human intervention. To be considered as an AV, a vehicle must be able to do most of the navigation-related tasks without human intervention. It must be able to drive itself to a place using roads that have not been specially designed for its purpose. Many famous and well-known companies are involved in developing and testing self-driving cars [27]. The AV is predicted to be the most disruptive transformation to human lives since the invention of internal combustion engines.

Self-driving cars with sophisticated electronics are now able to navigate and make judgments in real time while on the road. Self-driving cars promise safer and more effective ways of transportation, which could eventually reduce the number of fatalities brought about by human error. Research conducted by Google and the American National Highway Traffic Safety Administration (NHTSA) found that approximately 93% of traffic accidents are the result of human error. Some of these errors include blurred eyesight, deafness, and the repercussions of drunk driving.

1.4.1 The role of visible light in AVs

Visible light, captured through cameras and used in various ways, serves as a fundamental component in the perception, navigation, and decision-making processes of self-driving cars. It complements other sensor technologies to create a holistic understanding of the driving environment for these vehicles.

Visible light plays a significant role in the development and application of AVs through various technologies and applications:

- **Vision systems:** Cameras utilizing visible light are integral components of self-driving cars. They capture real-time visual data of the surroundings, assisting in object detection, lane recognition, traffic signs interpretation, and pedestrian identification. Image processing algorithms analyze this visual information to make driving decisions.
- **Lighting and signaling:** Visible light is crucial for traditional signaling and lighting systems on roads. Self-driving cars utilize these visual cues, such as

traffic lights, stop signs, and pedestrian crossings, to understand and respond to traffic regulations and ensure safe navigation.

- **Mapping and localization:** Visible light data captured by cameras aids in creating high-definition maps used for localization and navigation. These maps help AVs understand their position, plan routes, and recognize landmarks based on visible light-based features.
- **Gesture and communication:** In the future, visible light might be employed for communication between pedestrians, cyclists, and AVs. Hand signals, light-based cues, or interactive displays could facilitate communication to enhance safety and cooperation on the road.
- **Daytime driving:** Visible light cameras are highly effective during daylight conditions, providing clear and detailed images of the environment. They enable self-driving cars to operate efficiently in well-lit situations, leveraging natural light for navigation and perception.
- **Obstacle detection:** Cameras using visible light assist in detecting various obstacles on the road, including vehicles, pedestrians, cyclists, and debris. Algorithms analyze these visual inputs to make decisions about navigation and collision avoidance.

1.5 VLC/OWC communications to flying objects

OWC technology can become the core technology of the sixth generation (6G) communication system since it is less expensive, has a rapid transmission rate, and is not frequency band constrained like RF communication technology, which is widely utilized in 5G. Second, with significant technological and financial advancements, UAVs are progressively evolving into a mobile communication terminal for communication network access. It goes without saying that the seamless integration of the ground network and space network is a must for the future information network. It is expected that the planned 6G wireless technology will enable full worldwide wireless connection at any time and from any location [28]. The development of worldwide 6G communication systems depends heavily on how to create a low-cost, high-capacity communication system. One significant solution for 6G wireless systems may be OWC technology, which is based on UAVs.

VLC technology presents a revolutionary solution for communication with low earth orbit (LEO) satellites and UAVs because of its ability to transmit data at high speeds while maintaining security and efficiency. The ability of VLC to improve the functionality and connection of aerial and space networks is expected to grow as research advances, aiding in the creation of next-generation communication systems. UAVs can act as communication relays by sending data to satellites and then ground stations via OWC. This feature improves the overall efficiency of data return systems, especially in situations when the use of conventional communication channels may be restricted or unreliable [29]. UAVs can improve satellite communication effectiveness in many efficient fields. Real-time data transmission between UAVs, satellites, and ground stations is possible through the integration of UAVs with satellite communication systems. This enhances situational awareness and response by facilitating the quick distribution of vital data acquired by UAV sensors and cameras. Moreover, UAVs can serve as communication relays, sending information to neighboring satellites from their position. This increases the coverage area and effective range of satellite networks, enabling UAVs to operate beyond ground station line-of-sight while preserving connectivity. Also, authentication and encryption techniques can be used to secure satellite communication links. The integration of UAVs with secure satellite links results in data transmitted from the UAV to the satellite being shielded from interception or manipulation, hence augmenting the communication system's overall security.

1.5.1 VLC/OWC for UAV-enhanced IoV

VLC from ground vehicles to UAVs is an emerging technology that shows promise for enhancing connectivity in the IoV ecosystem. The use of VLC in this context offers several advantages, including high bandwidth capacity, reduced electromagnetic interference, and improved security compared to traditional RF communications. The major advantage is the predominant availability of line-of-sight (LoS) channel conditions, since the UAVs are up in the air.

In IoV applications, VLC can enable ground vehicles to transmit various types of data to UAVs, such as traffic information, vehicle status, or emergency alerts. This uplink communication can be particularly useful in scenarios where UAVs act as aerial relays or mobile edge computing nodes, facilitating data collection and distribution in smart transportation systems. The LoS nature of VLC also makes it suitable for precise localization and positioning of vehicles, which is crucial for applications like autonomous driving and traffic management.

However, implementing VLC from ground vehicles to UAVs faces several challenges. These include maintaining stable links despite the mobility of both vehicles and UAVs, mitigating the effects of ambient light interference, and ensuring reliable communication under various weather conditions. Additionally, the limited range of VLC compared to RF systems necessitates careful planning of UAV deployment and flight paths to maintain consistent coverage. Despite these challenges, ongoing research and development in this field is addressing these issues, making VLC a promising technology for enhancing connectivity and enabling new applications in the evolving landscape of IoV.

Continued research into improving link stability and performance in various atmospheric conditions Development of more efficient and compact optical communication payloads for UAVs Integration of OWC with other sensing and communication technologies for enhanced UAV capabilities. Overall, optical wireless communication and VLC with UAVs shows promise for enabling high-speed, long-range communication links in various applications, though challenges remain in optimizing performance and overcoming environmental factors.

1.5.2 VLC/OWC for UAV-LEO satellite communications

To improve data transmission capabilities, UAVs are increasingly being linked with optical wireless communication technologies, especially free-space optical (FSO) systems. UAVs can act as relay nodes in satellite communications, increasing connection and lessening the strain on satellite systems. This integration is important for a number of applications, including satellite communications. Compared to conventional RF communication systems, the combination of UAVs and optical wireless communication offers a number of benefits, including increased data throughput, decreased latency, and enhanced resilience against interference. In order to maximize performance in dynamic situations, recent research has concentrated on creating frameworks that take advantage of advanced modulation techniques including OFDM.

- **OFDM framework:** In an effort to improve wireless communication links between UAVs and ground stations, a novel framework combining OFDM with UAV-based FSO communication has been developed. This method overcomes the shortcomings of the current models by increasing data rates and adjusting to different air conditions and distances [30].
- **High altitude platforms (HAPs):** As a intermediary between satellites and ground terminals, UAVs can serve as HAPs. This configuration improves the total connection budget and reduces the processing load on satellites, resulting in a more effective communication system [29].
- **Channel modeling** Doppler effects and air turbulence have been taken into account while simulating the optical communication channels between UAVs and satellites in research. These models play a critical role in link parameter optimization and system performance prediction [31].
- **Experimental prototypes:** A number of experimental configurations have been created in order to evaluate the effectiveness of optical communication systems based on UAVs. These experiments have shown the possibility of high data rates and efficient transmission in a range of environmental settings [29].

Significant improvements in network infrastructures are anticipated as a result of the ongoing study in UAV-satellite OWCs, especially given the growing demand for high-speed data transmission. The evolution of next-generation wireless networks, particularly those that go beyond 5G, is expected to be significantly influenced by the integration of UAVs with optical technologies. In conclusion, a promising future for improving connectivity and communication efficiency, especially in satellite systems, is the synergy between UAVs and OWC technologies.

1.5.3 AI-enhanced UAV-satellite communications

AI can be used to improve the security of communication between UAVs and LEO satellites. This technology offers several key benefits and applications for businesses.

- **Secure data transmission:** The confidentiality and integrity of data transferred between drones and satellites are guaranteed by AI-enhanced drone–satellite communication security. Businesses can safeguard sensitive information from illegal access and eavesdropping by encrypting data and utilizing sophisticated authentication procedures.

- **Threat mitigation and detection:** AI systems are able to examine communication patterns, spot irregularities, and flag questionable activity. This helps companies avoid potential security breaches by blocking unauthorized access to drones or satellites and enabling them to notice and respond to threats in real time.
- **Enhanced situational awareness:** Businesses can see a whole picture of their satellite and drone operations thanks to AI-enhanced drone–satellite communication security. Businesses may improve situational awareness and facilitate well-informed decision-making by gaining insights into the whereabouts, status, and activities of their drones and satellites through the monitoring of communication channels and data analysis.
- **Compliance and regulatory adherence:** Companies using satellites and drones have to abide by a number of laws and industry norms. Businesses can comply with these regulations by using AI-enhanced drone–satellite communication security, which guarantees data transmission security and safeguards confidential information.
- **Enhanced operational efficiency:** AI-enhanced drone–satellite communication security boosts operational efficiency by automating security duties and offering real-time information. Companies may concentrate on their main goals while streamlining their security procedures and minimizing manual inputs.

Finally, companies using drones and satellites can benefit greatly from AI-enhanced drone–satellite communication security. Enterprises can function with confidence and safeguard their vital resources thanks to its improved data security, threat detection and mitigation capabilities, improved situational awareness, compliance assurance, and operational efficiency optimization.

1.6 VLC V2X research directions

VLC technology offers significant potential for high-speed data communication, but it faces several challenges. Achieving data rates higher than 1 Gbit s^{-1} remains difficult even with advanced signal processing techniques [32]. Additionally, integrating VLC with other wireless standards like RF is complex. Studies such as [33] have shown that combining high-speed VLC downlinks with lower-speed RF wireless uplink can improve latency and throughput. Also, the work in the OMEGA project aims to combine different RF and optical wireless to achieve the desired performance [34].

The IEEE 802.15.7 group has highlighted the potential of integrating lighting and communication for VLC technology [35]. They identified two main challenges: flicker mitigation and support for dimming. Techniques to address these challenges are defined in the IEEE 802.15.7 VLC standard. However, several technical challenges remain, such as understanding channel models for VLC, especially for non-line-of-sight (NLOS) outdoor environments. Networking light sources and upgrading current infrastructures to support communication also require support from the lighting industry.

Despite these challenges, VLC is being developed as a global industry standard in IEEE 802.15.7 and promises to be a future high data rate and power-efficient technology. However, using VLC in outdoor applications still faces many problems.

1.7 Chapter summary

This chapter introduces the potential use of VLC for vehicular applications, and studies AI based enhancements, especially within the realms of ITS. It begins with relevant background information and motivation and then describes various key aspects of VLC and OWC for the realization of IoV.

References

[1] Diller J and Fernando X 2022 A study of VLC between vehicles and traffic signal lights *Computer Networks, Big Data and IoT* ed A Pasumpon Pandian, X Fernando and W Haoxiang (Singapore: Springer Nature Singapore) pp 643–55

[2] Hassan N, Fernando X and Woungang I 2024 An emergency message routing protocol for improved congestion management in hybrid RF/VLC VANETs *Telecom* **5** 21–47

[3] Garcia M H C, Molina-Galan A, Boban M, Gozalvez J, Coll-Perales B, Şahin T and Kousaridas A 2021 A tutorial on 5G NR V2X communications *IEEE Commun. Surv. Tutor.* **23** 1972–2026

[4] Tong W, Hussain A, Bo W X and Maharjan S 2019 Artificial intelligence for vehicle-to-everything: a survey *IEEE Access* **7** 10823–43

[5] Singh G, Srivastava A and Bohara V A 2022 Visible light and reconfigurable intelligent surfaces for beyond 5G V2X communication networks at road intersections *IEEE Trans. Veh. Technol.* **71** 8137–51

[6] Kak A and Akyildiz I F 2021 Designing large-scale constellations for the internet of space things with cubesats *IEEE Internet Things J* **8** 1749–68

[7] World Health Organaization. 2024 Global transport ministers call for new road safety assessment framework for business available online: https://www.who.int/news/item/27-05-2024-global-transport-ministers-call-for-new-road-safety-assessment-framework-for-business

[8] Papadimitratos P, La Fortelle A D, Evenssen K, Brignolo R and Cosenza S 2009 Vehicular communication systems: enabling technologies, applications, and future outlook on intelligent transportation *IEEE Commun. Mag.* **47** 84–95

[9] Kumar N 2011 Visible light communication systems for road safety applications *PhD Thesis* Universidade de Aveiro

[10] Agarwal A and Little T D C 2010 Role of directional wireless communication in vehicular networks *2010 IEEE Intelligent Vehicles Symp. (June 2010)* pp 688–93

[11] Little T D C, Agarwal A, Chau J, Figueroa M, Ganick A, Lobo J, Rich T and Schimitsch P 2010 Directional communication system for short-range vehicular communications *2010 IEEE Vehicular Networking Conf.* pp 231–8

[12] Liu C B, Sadeghi B and Knightly E W 2011 Enabling vehicular visible light communication (V2LC) networks *Proc. Eighth ACM Int. Workshop on Vehicular Inter-networking (VANET'11)*, ACM pp 41–50

[13] Chan C H, Liu H, Kwan T and Pang G K H 1998 Dual use of LEDs: signaling and communications in ITS *Proc. 5th World Congress on Intelligent Transport Systems, Paper 3035 (Seoul, Korea)* pp 12–6
[14] Liu H S and Pang G 2003 Positioning beacon system using digital camera and LEDs *IEEE Trans. Veh. Technol.* **52** 406–19
[15] Wada M, Yendo T, Fujii T and Tanimoto M 2005 Road-to-vehicle communication using LED traffic light *IEEE Proc. Intelligent Vehicles Symp., 2005* pp 601–6
[16] Hara T, Iwasaki S, Yendo T, Fujii T and Tanimoto M 2007 A new receiving system of visible light communication for ITS *2007 IEEE Intelligent Vehicles Symp.* pp 474–9
[17] Nagura T, Yamazato T, Katayama M, Yendo T, Fujii T and Okada H 2010 Improved decoding methods of visible light communication system for ITS using LED array and high-speed camera *2010 IEEE 71st Vehicular Technology Conf.* pp 1–5
[18] Yamazato T *et al* 2014 Image-sensor-based visible light communication for automotive applications *IEEE Commun. Mag.* **52** 88–97
[19] Terra D, Kumar N, Lourenço N, Alves L N and Aguiar R L 2011 Design, development and performance analysis of DSSS-based transceiver for VLC *2011 IEEE EUROCON—Int. Conf. on Computer as a Tool* pp 1–4
[20] Kumar N, Lourenço N, Terra D, Alves L N and Aguiar R L 2012 Visible light communications in intelligent transportation systems *2012 IEEE Intelligent Vehicles Symp.* pp 748–53
[21] Khan L U, Yaqoob I, Imran M, Han Z and Hong C S 2020 6G wireless systems: a vision, architectural elements, and future directions *IEEE Access* **8** 147029–44
[22] Bolsunovskaya M, Leksashov A, Shirokova S and Tsygan V 2021 Development of an information system structure for photo-video recording of traffic violations *E3S Web Conf.* **244** 07007
[23] Li J, Cheng H, Guo H and Qiu S 2018 Survey on artificial intelligence for vehicles *Automot. Innov.* **1** 2–14
[24] Farahneh H, Hussain F and Fernando X 2018 Performance analysis of adaptive OFDM modulation scheme in VLC vehicular communication network in realistic noise environment *EURASIP J. Wirel. Commun. Netw.* **2018** 243
[25] Soner B and Coleri S 2021 Visible light communication based vehicle localization for collision avoidance and platooning *IEEE Trans. Veh. Technol.* **70** 2167–80
[26] Filiz C 2020 Can autonomous vehicles prevent traffic accidents? *Accident Analysis and Prevention* ed M Darçın (London: IntechOpen)
[27] Namatherdhala B, Mazher N and Sriram G K 2022 Uses of artificial intelligence in autonomous driving and V2X communication *Int. Res. J. Modern. Eng. Technol. Sci.* **4** 1932–6
[28] Saad W, Bennis M and Chen M 2020 A vision of 6G wireless systems: applications, trends, technologies, and open research problems *IEEE Netw.* **34** 134–42
[29] Ding J, Mei H, Chih-Lin I, Zhang H and Liu W 2020 Frontier progress of unmanned aerial vehicles optical wireless technologies *Sensors* **20** 5476
[30] Elsayed E E 2024 Investigations on OFDM UAV-based free-space optical transmission system with scintillation mitigation for optical wireless communication-to-ground links in atmospheric turbulence *Opt. Quantum Electron.* **56** 837
[31] Li M, Hong Y, Zeng C, Song Y and Zhang X 2018 Investigation on the UAV-to-satellite optical communication systems *IEEE J. Sel. Areas Commun.* **36** 2128–38

[32] O'Brien D C, Zeng L, Le-Minh H, Faulkner G, Walewski J W and Randel S 2008 Visible light communications: challenges and possibilities *2008 IEEE 19th Int. Symp. on Personal, Indoor and Mobile Radio Communications* pp 1–5

[33] Hou J and O'Brien D C 2006 Vertical handover-decision-making algorithm using fuzzy logic for the integrated radio-and-OW system *IEEE Trans. Wirel. Commun.* **5** 176–85

[34] OMEGA. Omega project available online: http://www.ict-omega.eu

[35] Rajagopal S, Roberts R D and Lim S K 2012 IEEE 802.15.7 visible light communication: modulation schemes and dimming support *IEEE Commun. Mag.* **50** 72–82

IOP Publishing

Visible Light Communications (Second Edition)
Vehicular applications
Xavier Fernando and Hasan Farahneh

Chapter 2

Nuts and bolts of V2X VLC systems

This chapter covers the fundamentals of visible light communication (VLC) systems in vehicle-to-everything (V2X) applications. It discusses the key components like VLC emitters and receivers, emphasizing the modulation techniques used to transmit data via light signals. The chapter analyzes the performance factors, including LED characteristics and noise interference, comparing VLC systems to traditional radio frequency (RF) communications. Safety concerns and potential collision avoidance systems are examined, alongside an example algorithm design for such systems. The chapter highlights VLC's dual functionality in illumination and communication, noting its significance in enhancing vehicular safety and connectivity in modern transportation infrastructure.

2.1 VLC basics

A VLC includes a transmitter, equipped with LEDs or laser diodes, to modulate light according to the provided information signal, which is often binary. It also has a receiver with photodetectors to detect variations in the received light and then extract the data. For a VLC channel to be established, typically, the transmitter and receiver must be aligned with each other. Figure 2.1 illustrates the basic architecture of a VLC system.

2.1.1 Visible light terminology

In this section, we will present some of the basic light terminologies.

- **The candela (cd):** This is the SI base unit of luminous intensity of a source that emits monochromatic radiation of frequency 540×10^{12} Hz in a given direction. That source will have a radiant intensity of 1/683 watt per steradian in that direction.
- **Luminance (cd m^{-2}):** A measure of the brightness of a surface, indicating the amount of light emitted or reflected in a given direction. It is measured in candelas per square meter.

doi:10.1088/978-0-7503-6049-4ch2

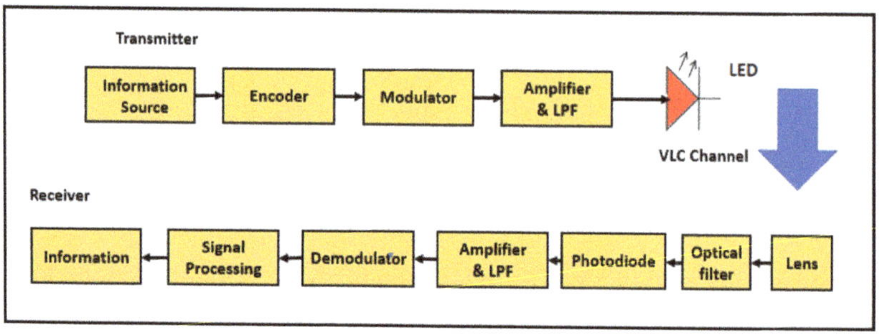

Figure 2.1. VLC transceiver block diagram.

- **Lux (lx):** An SI derived unit of illuminance (luminous intensity), measuring the amount of light (luminous flux) falling on a unit area.
  1 lx = 1 lumen per square meter.
  The Lux can be converted to watts m^{-2} for daylight by multiplying it with the number 0.004 02. This conversion is only appropriate for the visible light band of interest [1].
- **Luminous efficacy of radiation (LER):** A measure of how efficiently a light source is able to convert electrical power (W) to light (lumens) and is expressed in lumens per watt (lm W^{-1}).
- **Solar insolation:** This is a measure of solar irradiance over of period of time—typically over the period of a single day.

The relation between lx and the power is given by [2]

$$1\ (\text{lux}) = 1\ (\text{lumens})/\text{m}^2 = \frac{\text{Power (W)} \times \text{LER}}{\text{Area (m}^2\text{)}} \tag{2.1}$$

Figure 2.2 illustrates the relationship between the illuminance and the candela [2].

2.2 VLC emitter

A VLC emitter is a device that transforms (binary) data into messages for transmission via a free-space optical medium using visible light. It serves a dual purpose: emitting light and transmitting data simultaneously with the same optical power, without noticeable flickering[1] or dimming[2] [3, 4]. The core of the VLC emitter is the encoder, which modulates the data into a message. This encoder manages the LED/laser's switching in accordance with the binary data being transmitted at a specific data rate, thereby converting the binary data into an appropriately modulated light beam.

The system in which the *intensity* of the light beam is modulated and the power is detected is called the intensity modulation and direct detection (IM/DD) system.

[1] Using high-frequency modulation that is beyond human perception (above 100 Hz) prevents visible flickering.

[2] Using advanced adaptive modulation techniques such as OFDM can dynamically adjust to maintain consistent light levels based on ambient light conditions and required data rates, ensuring that visible light levels remain constant.

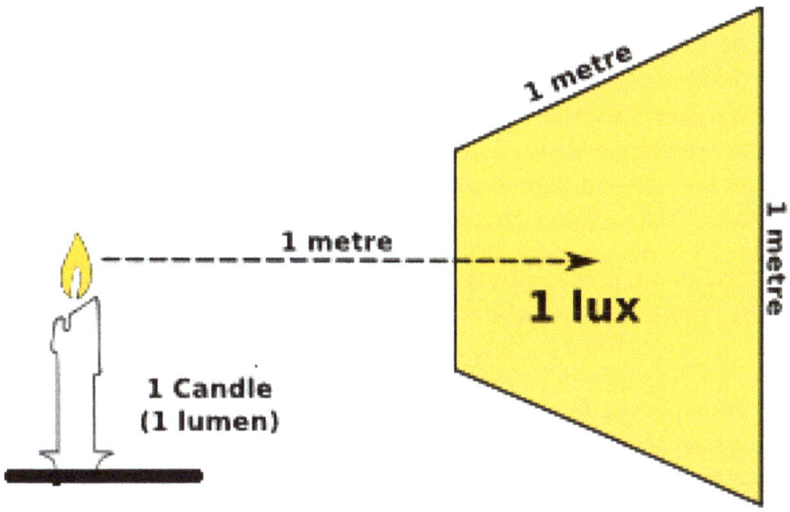

Figure 2.2. The relationship between the illuminance and the candela [2].

The simplest scheme is the ON-OFF-Keying (OOK) amplitude modulation. However, due to the susceptibility of amplitude to noise interference, techniques independent of pulse amplitude like pulse width modulation (PWM) or pulse position modulation (PPM) are also commonly employed in optical wireless communication (OWC) systems. More advanced modulation methods such as orthogonal frequency division multiplexing (OFDM) [5], discrete multi-tone modulation (DMT) or DSSS discrete sequence spread modulation can be utilized [6], offering further benefits albeit at greater complexity.

It should be noted that the systems, theories, and techniques outlined in this book are relevant to any OWC system, regardless of the operational wavelength. This includes infrared, single wavelength sources, and broadband transmitters, which are all compatible with our channel models, coding, and equalization analysis. Nevertheless, the focus will be more on visible light sources, as they align with the scope of this book and have the dual functionality of illumination and communication.

VLC typically employs white light LEDs used in lighting. Since a PN junction emits only a single wavelength, white light generation using solid-state technology has only recently matured. Presently, there are two main types of white light emitters: RGB LED and phosphorescent LED. An RGB LED combines red, green, and blue chips, mixing the three colors in the correct proportion to produce white light. RGB LEDs have a relatively higher modulation bandwidth, up to approximately 20 MHz, and can support wave division multiplexing (WDM), but they are comparatively costly. On the other hand, phosphorescent LEDs use a blue LED chip coated with yellow phosphor, which is the most widely used white LED due to its affordability. However, the phosphor's slow response time limits the modulation bandwidth of these LEDs to just a few MHz. Phosphor-based LEDs have turn-on and turn-off times ranging from tens to hundreds of nanoseconds, significantly slower than direct emission LEDs.

The performance parameters of a VLC emitter are predominantly determined by the LED's characteristics. The data transmission rate of the VLC system hinges on the LEDs' ability to switch rapidly, while the size of the emitter influences its transmission power. LEDs designed for direct emission usually activate within a few nanoseconds, with larger LEDs requiring slightly more time. Deactivation times are generally in the tens of nanoseconds range, which is somewhat slower than activation times. This discrepancy is due to the fact that standard LED driving configurations do not actively decrease the LED's voltage during deactivation. Infrared LEDs are often the quickest to transition, for reasons that will be discussed subsequently. There are also specially engineered LEDs that are optimized for ultra-fast switching, capable of handling pulses ranging from 800 ps to 2 ns.

In addition to the inherent turn-on and turn-off times, the following factors also effect the switching speed:

- Inductance of the wires or copper traces causes longer rise and fall times. The longer the traces, slower the transitions.
- Junction capacitance of the LED/laser will impact its switching speed. The junction capacitance is directly proportional to the radiating area of the LED.
- Parasitic capacitance (traces and support circuitry) plays an important role in increasing the RC time constant and thus slowing transitions.

Due to the inductive and capacitive elements mentioned, an LED with a higher forward voltage will exhibit longer rise and fall times, necessitating greater driving currents. Consequently, infrared LEDs, which generally have the lowest forward voltages, have the quickest transitions. Typically, the limiting time constants are in the range of hundreds of nanoseconds, primarily influenced by external factors such as the driving circuit, as opposed to the LED junction's significantly shorter transition times.

2.2.1 LED or laser transmitter

Although LEDs are predominantly described in this book, laser transmitters can also be used in VLC. Note laser headlights have been increasingly deployed in high-end vehicles and they can produce 100s of times the intensity of LEDs and reach distances of up to 600 m, whereas LED headlights typically reach distances of about 300 m. Also, laser headlights Consume only half as much power as LEDs while producing more intense light. While controlling LED beams require matrix (array) designs, laser beams can be precisely controlled. However, laser headlights are currently very expensive and limited to high-end vehicles, whereas LED headlights are more widely adapted. Both technologies are designed to be safe for road use, but the higher intensity of laser headlights may require additional safety measures for VLC applications (table 2.1).

- **Modulation:** Both lasers and LEDs can be rapidly modulated, which is crucial for data transmission in VLC. However, some laser systems may offer higher modulation speeds.

Table 2.1. LED and laser as VLC transmitters.

	LASER	LED
Divergence	Laser light can be highly collimated, resulting in a very narrow beam. The divergence of a laser beam can be extremely small, often measured in milliradians (mrad)	LEDs have much higher beam divergence. They emit light in a wider angle, more uniformly in all directions. This makes it challenging to create a narrow, focused beam with LEDs without additional optics.
Range	Due to their low divergence and high intensity, laser headlights can reach distances of up to 600 m. This longer range makes lasers potentially advantageous for long-distance VLC applications.	LED headlights typically reach distances of about 300 m. While this is less than lasers, it's still sufficient for many VLC applications, especially in urban environments or shorter-range communications.
Light Intensity	Lasers can produce extremely high light intensities. This high intensity, combined with low divergence, allows for long-range transmission of data in VLC systems.	While not as intense as lasers, LEDs can still produce significant light output. Their broader emission pattern may be advantageous for some VLC applications where wide-area coverage is needed rather than a narrow beam.

- **Cost and complexity:** LED systems are generally less expensive and less complex than laser systems. This makes LEDs more accessible for widespread VLC applications.
- **Safety:** The high intensity of laser light raises safety concerns, particularly for eye safety. LED systems generally have fewer safety issues due to their lower intensity and higher divergence.
- **Power efficiency:** Laser headlights are reported to consume only half as much power as LEDs while producing more intense light. This could be advantageous for energy efficient VLC systems.

In conclusion, while lasers offer advantages in terms of range, intensity, and low divergence for VLC applications, LEDs provide a more cost-effective and safer option with sufficient performance for many use cases. The choice between LED and laser for VLC would depend on the specific requirements of the application, including range, data rate, cost constraints, and safety considerations.

2.3 VLC receiver

The VLC receiver will be detecting the data from the received modulated light beam using an appropriate photodiode (PD). It transforms the light into an electrical signal that will be demodulated and decoded by the embedded decoder module. Usually, the PD has more modulation bandwidth than LED and the

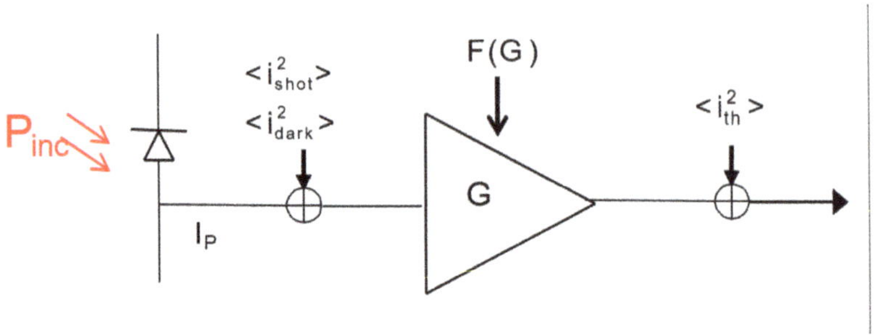

Figure 2.3. Signal and noise elements of a generalized photodiode.

signal-to-noise-ratio (SNR) at the receiver is the primary design constraint. Optical filtering is often applied at the receiver to reduce out-of-band noise [7].

There are two widely adopted types of PDs, ordinary positive intrinsic negative (PIN) photodiode and avalanche photodiode (APD). Major signal and noise elements of a generalized PD is shown in figure 2.3. Here G is the avalanche gain and $F(G)$ is the excess noise of the APD. For PIN diodes, we may assume $G = F(G) = 1$ and use the same diagram without losing the generality. The dark current noise, $\langle I_{\text{dark}}^2 \rangle$ is usually negligible in VLC environments. Thermal noise $\langle I_{\text{th}}^2 \rangle$ arises from the bias resistance and it is a constant for a given receiver. It does not change with the received signal. The shot or quantum noise, $\langle I_{\text{shot}}^2 \rangle$ is proportional to the received optical power. Note, the received signal power includes both the energy transmitted from the OWC transmitter and from ambient light. Often the ambient light can be of the same order or even larger than the transmitted signal. Hence, the quantum noise can be very large and it will often dictate the performance of an OWC receiver.

Figure 2.4 shows the variation of quantum and thermal noise powers as a function of received signal (plus in-band ambient noise). As depicted in figure 2.4, the APD is advantageous when dealing with weak signal strengths. The APD's internal gain multiplies both the received signal and the associated shot noise prior to the addition of thermal noise. It's important to note that quantum noise is directly proportional to the sum of the received signal and ambient light power within the detection bandwidth. In contrast to fiber optic systems, OWC receivers typically encounter substantial in-band ambient light, leading to significant shot noise. This amplified quantum noise can drastically degrade the SNR. Consequently, despite its inherent gain, the APD is generally not preferred in OWC systems due to its amplification of ambient noise-induced quantum noise.

The key parameter in PDs is the responsivity γ, which is commonly defined as,

$$\gamma = \frac{I_{\text{p}}}{P_{\text{inc}}} \tag{2.2}$$

where I_{p} denotes the average generated photocurrent, and P_{inc} denotes the incident optical power.

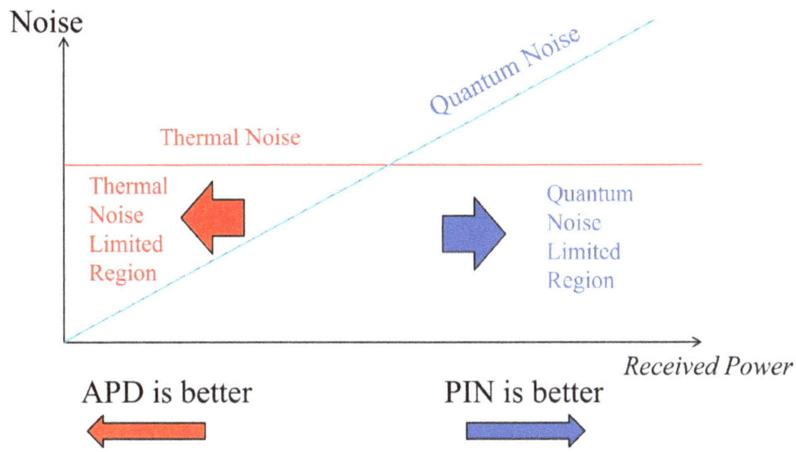

Figure 2.4. Depiction of quantum and thermal noise impacts in a photodetector. The APD performs better when the received signal is very small because the quantum noise will also be small. However, the PIN diode performs better when the quantum noise is large and the limiting factor.

The PD responsivity depends on the physical structure of the PD and has the units of (A W^{-1}). The incident power P_{inc} is proportional to its effective light collection area.

The effective area of the PD must be sufficiently large to gather adequate transmitted signal power. However, a larger area increases the junction capacitance, which can slow down the PD. Speed is also affected by the electron transit time, which is the duration required for electrons to traverse the depletion layer. A wider depletion layer leads to a longer transit time, but it also absorbs more photons, enhancing responsivity. Thus, designing a PD for OWC applications significantly differs from designing one for fiber optic applications, where detectors have a very small photosensitive area and are capable of operating at gigabit speeds, whereas OWC detectors have a larger photosensitive area. Therefore, their speed is usually lower.

It is important to note that the incident light on the receiver comes not only from the emitter but also from other light sources, both artificial and natural, leading to considerable interferences. The performance of the VLC receiver can be improved with an optical filter that eliminates the unwanted spectral components of the interfering noise signal [7]. The SNR can also be improved by adaptively controlling the field of view (FOV) of the receiver to minimize ambient noise while maximizing received signal strength, as elaborated later in the book. Overall, the selection of a VLC receiver should balance cost, performance, and sensitivity to meet the specific requirements [8].

In OWC systems, the most important challenge is the background noise which is created from ambient light in the environment. Ambient light can rise from sunlight, skylight, advertising signs and fluorescent lamps, or other light sources [9]. The background noise existing in the environment leads to a constant (DC) photocurrent that causes shot noise at the receiver. This noise source is signal independent and is the result of the summation of many independent Poisson distributed random variables. Hence, it can be modeled as a Gaussian distribution [8].

The shot (quantum) noise variance is given by [10]

$$\sigma^2_{\text{Shot}} = 2\, q_{\text{e}}\, G^2\, F\, (I_{\text{inc}})\, B \tag{2.3}$$

where F denotes the excess noise, q_{e} denotes the electronic charge constant ($1.6 \times 10^{(-19)}$ coulomb), G is the PD gain, B denotes the electrical bandwidth of the PD, and I_{inc} is the photocurrent produced due to the received optical power.

The photocurrent I_{inc} is given by [11],

$$I_{\text{inc}} = G\, A_{\text{r}} \int_{\lambda_1}^{\lambda_2} P_{\text{inc}}(\lambda)\, \gamma(\lambda)\, T_{\text{o}}(\lambda)\, d\lambda \tag{2.4}$$

where, λ_1, λ_2 are the visible light wavelengths, $\gamma(\lambda)$ is the responsivity of the APD given in A W^{-1} at certain λ, A_{r} is the effective area of the receiver, and $T_{\text{o}}(\lambda)$ is the transmittance of the bandpass optical blue filter.

Although the shot (quantum) noise is dominant, there are other types of noises that are liable for degrading the performance of the VLC system, such as dark noise and thermal noise. Dark noise is caused due to the avalanche effect of the APD and is given by [10]

$$\sigma^2_{\text{Dark}} = 2\, q_{\text{e}}\, G^2\, F\, I_{\text{dg}}\, B + 2\, q_{\text{e}}\, I_{\text{ds}} \tag{2.5}$$

where I_{ds} is a surface dark current, and I_{dg} is a bulk dark current.

Also, due to a random thermal motion of charge carriers, thermal noise is generated. The variance of thermal noise is given by [10]

$$\sigma^2_{\text{Thermal}} = 4 \left(\frac{K_{\text{B}}\, T_k}{R_L} \right) F_n\, B \tag{2.6}$$

where K_{B} denotes Boltzmann's constant, T_k denotes absolute temperature, R_L denotes the load resistance, and F_n is the photodiode noise figure.

To find the variance of the total noise, we have to apply variance sum law which is an expression for the variance of the sum of two variables. If the variables are independent, the following formula represents the variance of the sum of the variables X and Y ($\sigma^2_{X+Y} = \sigma^2_X + \sigma^2_Y$).

If X and Y are correlated, then the variance sum is ($\sigma^2_{X+Y} = \sigma^2_X + \sigma^2_Y + 2\, \rho_{xy}\, \sigma_X\, \sigma_Y$), where ρ_{xy} is the correlation coefficient between X and Y [12].

Therefore, the variance of the total noise σ^2_n is given by,

$$\sigma^2_n = \sigma^2_{\text{shot}} + \sigma^2_{\text{Dark}} + \sigma^2_{\text{Thermal}} \tag{2.7}$$

Table 2.2 shows the probability density function (PDF) of each type of noise.

Table 2.2. PDFs of different types of noise.

Noise	Probability density function
Shot noise	Poisson distribution [13]
Dark noise	Poisson distribution [13]
Thermal noise	Gaussian distribution [13]
ISI noise	Gaussian distribution [14]
Clipping noise	Gaussian distribution [15]

2.3.1 Image sensors as VLC receivers

Many modern vehicles are equipped with various forms of image sensors such as rear-view and side-view (rolling shutter) cameras to enable precise navigation. There are interesting proposals to use these image sensors as VLC receivers with a dual purpose. Using image sensors as VLC receivers offers the advantage of high spatial resolution and the ability to capture multiple signals simultaneously, making them suitable for complex environments and integrated applications. Moreover, the camera receiver provides the non-interference communication capability due to the excellent spatial separation capability of the image sensor mounted in the camera. Therefore, non crosstalk communication with multi-LEDs without a complicated protocol and processing is achieved by using a camera based image sensor. It also prevents optical signals from being mixed with noise (incident sunlight) and enables simple link designs.

Rolling shutter is a method of image acquisition in which each frame is recorded not from a snapshot obtained at a single point of time, but rather by scanning across either vertically or horizontally [16]. The transmitting LED light switches on and off at very high frequencies according to the modulation, and the pixels of the camera sensor activate sequentially (by row), but does not get the entire image simultaneously. When the rows of pixels are activated, they are exposed to the light at that time and then their values are stored. After the procedure is completed, the scan lines captured at different times are merged together to form a single image.

However, they come with trade-offs in terms of lower data rates, higher power consumption, and the need for complex signal processing. Balancing these pros and cons is crucial when considering image sensors for specific VLC applications. Each type of receiver has its strengths and limitations, and the optimal choice will depend on factors such as required data rate, range, complexity, and cost constraints of the VLC system. Table 2.3 a comparison of APDs, PIN photodiodes, and image sensors as VLC receivers.

In this book, Hamamatsu-S8664 APD is considered as a receiver. The simulation parameters and their values are given in table 2.4 [17].

Table 2.3. Comparison of APDs, PIN photodiodes, and image sensors as VLC receivers.

	APD	PIN	Image Sensor
Sensitivity and gain	Offer high sensitivity due to internal gain, hence suitable for low-light detection. However, this could be a drawback when the ambient light noise is high.	Lower sensitivity than APDs since, no internal gain mechanism.	Typically use PIN photodiode structures, so sensitivity is similar to individual PIN photodiodes.
Speed	Fast response times, suitable for high-speed applications up to GHz range.	Fast response times, suitable for high-speed applications up to GHz range.	Frame rate limited, usually slower than individual photodiodes for VLC applications.
Noise	Higher noise due to avalanche process, but can offer better SNR when the quantum noise is smaller than the thermal noise.	Better SNR than APDs when the quantum noise is bigger than the thermal noise.	Noise characteristics similar to PIN photodiodes, but may have additional sources of noise from readout circuitry.
Spectral response	Available in various materials (e.g., silicon, InGaAs) to cover different spectral ranges.	Available in various materials (e.g., silicon, InGaAs) to cover different spectral ranges.	Typically silicon-based, covering visible to near-infrared range.
Area capability	Can be arranged in arrays, but typically smaller than image sensors.	Can be arranged in arrays, but typically smaller than image sensors.	Naturally arranged in large 2D arrays, allowing for spatial diversity and imaging capabilities.
Cost and complexity	Generally more expensive and require more complex biasing and control circuitry.	Simpler and less expensive than APDs.	More complex readout circuitry, but benefit from mass production for consumer electronics.
VLC-specific considerations	Excellent for long-range or *low-light* VLC applications due to high sensitivity.	Suitable for most VLC applications, offering a good balance of performance and cost.	Can provide spatial diversity and allow for multiple transmitter tracking, but may be limited in speed for high-data-rate VLC.

Table 2.4. Noise parameters and their simulation values.

Symbol	Description	Value
APD model	(Si) Hamamatsu	S 8664 – 1010
λ	Visible light wavelength	350–750 nm
G	APD gain	50
q_e	Electron charge	1.9×10^{-19} C
B	APD bandwidth	65 MHz
I_{ds}	Surface dark current	10 nA
I_{dg}	Bulk dark current	5100 μA
K_B	Boltzmann's constant	1.38×10^{-23} J K^{-1}
T_k	Temperature	298 K
F_n	APD noise figure	0.2
γ	Photodetector responsivity	0.28 A W^{-1}
R_L	Load resistance	50 Ω

2.4 VLC channel

The VLC channel represents the unobstructed space or air path between the transmitter and receiver. At the receiver end, the power intensity of the received signal diminishes proportionally to the square of the distance from the transmitter, based on the free-space propagation model. Consequently, the received signal is typically weak. Moreover, the VLC channel may be affected by various sources of optical noise. The Sun is the predominant source of noise during the day. The unpredictability of the VLC channel is due to the different noise sources and varying weather conditions. Environmental factors such as rain, snow, dust, or dense fog can degrade the quality of the VLC link by scattering the light that carries the data. These noise sources, coupled with the inherently low signal power at greater distances, significantly impact the SNR in VLC systems.

Another key feature of the VLC channel is the directional propagation of light, which makes the line-of-sight (LoS) channel predominant. The multipath, or reflected signals, collected at the receiver have a minimal impact and are only noticeable at short distances between the emitter and receiver.

The angular alignment between the transmitter and receiver, as well as their FOVs, also plays a crucial role in the SNR. The SNR is at its peak when the receiver is perpendicular to the transmitter and diminishes as the angle deviates from this position.

The SNR can be improved if the receiver 'looks away' from the Sun's rays. However, this has to be done without reducing the received signal power, which is a 3D tracking problem, where both transmitter and receiver are moving. There have been a few research approaches to solve this issue. For example, authors of [18] analyzed the resulting optimization problem and found the optimal angle at which the instantaneous SNR is maximized using an extended-Kalman filter. This depends on the incident angle of the optical signal as well as the incident angle of the sunlight.

The above-mentioned solution is more difficult when there is a significant elevation difference between the transmitter and the receiver, like when traffic signal lights communicate with the waiting vehicles via VLC. However, due to the directional properties of light propagation and the height differences between the vehicle and traffic signal, there are stringent limitations on a feasible range of communications. In [19], this problem is studied using a geometrical approach. It proposes a new method for improving performance by segmenting the transmitter to create an array of axially misaligned sources focused at different spatial regions.

2.5 Comparison of VLC and RF-based V2X communications

Vehicular communication systems play a crucial role in enhancing road safety, traffic management, and overall driving experience. Both RF-based communication and VLC have their unique advantages and challenges when applied to vehicular environments (table 2.5). Comprehending their major properties is crucial for assessing the advantages and limitations of each technology. Here is a detailed comparison of the two.

In summary, both RF-based and VLC-based vehicular communications have unique advantages and challenges. RF-based communication offers broader coverage and robust infrastructure but faces interference and security challenges. VLC-based communication provides high-speed, secure, and interference-free data transmission but requires clear LoS and is more suitable for short-range applications. The choice between the two technologies depends on the specific requirements and constraints of the vehicular communication application in question.

Table 2.5. Comparison of RF and VLC technologies for vehicular communications.

Aspect	RF-based communication	VLC-based communication
Technology	Cellular (4G, 5G), DSRC, WAVE, Wi-Fi	Generic VLC, Li-Fi, IEEE 802.11bb™
Frequency	GHz (e.g., 2.4 GHz, 5.9 GHz)	Visible light spectrum (400–800 THz)
Data Speed	High (up to several Gbps)	Very high (potentially up to 100 Gbps)
Latency	Low (1 ms in 5G)	Very Low, depends on transceiver electronics.
Range	Long range, omnidirectional	Short-range (LoS), directional, affected by obstructions
Interference	Susceptible to RF interference	Immune to RF interference, but affected by ambient light, fog rain etc
Security	Requires encryption, vulnerable to RF attacks	More secure due to limited range and LoS requirements
Cost	High due to infrastructure needs	Lower, especially if integrated with existing LED systems
Applications	Mission critical as well as other tasks	Short-range, high-speed tasks

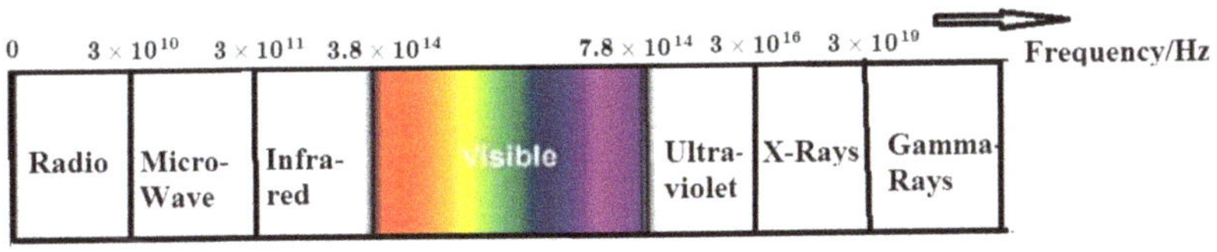

Figure 2.5. Distribution of the electromagnetic spectrum. Reproduced with permission from [20].

2.6 VLC advantages and drawbacks

VLC offers several unique advantages and a few drawbacks. One key advantage is its high bandwidth capacity, as the visible light spectrum is significantly larger than the RF spectrum, enabling potentially higher data transmission rates and supporting high-throughput applications, as shown in figure 2.5. VLC uses unlicensed spectrum, and it is safe for humans and for high-precision electronic equipment. VLC also enhances data security because it requires a direct LoS, making it less susceptible to eavesdropping and unauthorized access. Additionally, VLC is immune to RF interference, making it suitable for environments with high RF congestion or where RF communication is restricted, such as hospitals and airplanes. It can also be seamlessly integrated with existing LED lighting infrastructure, providing a cost-effective solution for data transmission. Moreover, VLC is a relatively green wireless communication technology [20].

However, VLC has notable drawbacks. The requirement for a direct LoS means that any physical obstruction can disrupt the communication link, limiting its reliability and range. Environmental factors like sunlight, fog, and dust can further degrade performance. Moreover, VLC cannot penetrate opaque objects, reducing its effectiveness in non-LoS scenarios. While VLC offers high data rates, these are typically confined to short-range communication. Lastly, retro-fitting existing lighting systems with VLC technology may involve significant upfront costs and technical challenges. Despite these drawbacks, VLC remains a promising technology for specific applications where its advantages can be fully leveraged.

2.7 Safety concerns of OWC systems

Infrared, visible, or ultraviolet electromagnetic radiation, in sufficient quantity or concentration, can cause damage to the human eye. Many cases of laser-induced eye damage have been documented. Visible light is usually safe as unknown exposure rarely happens (we see the light and naturally close our eyes if it is too bright). On the other hand, infrared (IR) cannot be seen and can damage the retina of the human eye. The amount of power that can be emitted by an IR radiator is controlled by several regulating bodies. At high intensities, it may even damage our skin. IEC 825-1 (Compliance of Infrared Communication Products) is the dominant standard that regulates the MPE (maximum permissible exposure), as a function of source area, exposure duration, and wavelength. Typically, MPE is higher at longer wavelengths with a bigger source area[3]. Note that the eye safety measures have to be strictly adhered to when *laser* transmitters are involved, even in VLC.

The initial standard, IEC 825, was first published in 1984 and based on the American National Standard ANSI Z-136.1. The most significant change that took place with the second edition of the standard (i.e., IEC 825-1) has been to include LEDs since recently the power emission capabilities of LEDs have significantly increased.

In essence, IR and VLC both involve considerations related to eye and skin safety, with VLC generally posing less risk due to its operation within the visible spectrum and adherence to IEC-825 guidelines. Note that, although generally ignored, RF communication also involves non-ionizing radiation, with safety guidelines focused on preventing thermal effects and managing potential long-term biological risks. By adhering to relevant safety standards and guidelines, the risks associated with each communication technology can be effectively managed to ensure safe usage.

2.8 An example collision avoidance system

This section's main objective is to suggest a system that will help drivers prevent collisions of any kind. The anti-collision system is designed to prevent accidents on the road by identifying approaching objects and classifying them accordingly, allowing the vehicle to decide if they qualify as obstacles or not. In this system, the vehicles will communicate among each other in terms of VLC, and by help of machine learning (ML) to predict possible accidents with barriers.

There are several collision avoidance technologies available, but they are all dependent on communication between automobiles or between automobiles and infrastructure. We recognize the accuracy with which these systems operate, but we also recognize their interdependence. In this section, we aim to develop an independent collision avoidance system. Since this device is going to be installed in cars owned by regular people, there is no need to modify anything about the vehicle. If there's a chance of colliding with another obstacle, the device will work and give the driver information. This system is more trustworthy and cost-effective because it doesn't depend on anyone else to work.

Moreover, the suggested system has an ultrasonic sensor and camera mounted above the car that aims to prevent traffic accidents, errors, and casualties. The camera records activity as it approaches the car and keeps an eye on everything the driver sees. On the other hand, the vehicle's mounted ultrasonic sensor begins to detect the separation between the car and the approaching object. Following its decision, it periodically transmits data to the processing unit so that it can perform the function in accordance with the object's distance from the vehicle as detected by the sensor.

The processing unit, which is built with internet of things (IoT) devices and uses networking technologies in a vehicle, receives input from the specified source (the

---

[3] MPE is approximately 1/10th of the level of electromagnetic radiation exposure, which will cause damage for 50% of human eyes.

vehicle's camera and ultrasonic sensor, which are installed on both sides) and begins interacting with the pattern recognition layer, which uses a weighted dataset to classify the input image pattern. Once the acquired images are classified into an obstacle category, the display unit receives a warning notification. This allows the driver to be informed when an obstacle is approaching their vehicle from a specific side. On the other hand, the processing unit analyzes the data from the sensor unit, such as the distance from the vehicle to the other vehicle approaching from the side, and using applied logic, it begins to determine whether or not the approaching object is expected to be dangerous. The warning notice will begin to be delivered to the vehicle if it is determined to be dangerous in nature.

In addition, the camera in front of the vehicle analyzes the traffic light that the car is approaching and continuously transmits its color to the display mounted in the driver's cabin of the car, enabling the driver to easily obtain information about the traffic light. The ML algorithm by use of datasets has been used to train the digital footprints of all traffic sign boards and traffic lights for accurate recognition. We are able to locate precise and up-to-date events occurring around the car by using cameras and ultrasonic sensors. The goal of this project is to demonstrate how IoT and ML approaches may work together. While ML with a camera refers to digital visuality for interpreting sign boards, traffic signals, and other things, IoT refers to actual sensors for the world surrounding the vehicle. After the sensors have acquired their data, a single core intelligence will compile all of the data and make the appropriate decisions to transmit the appropriate notification to the drivers' display. The car's sound and display modes, which monitor a variety of circumstances surrounding the vehicle, will be used to communicate decisions and alerts to the driver. The driver-site display device in the car that is part of the system is not very user-friendly. But the driver may readily monitor activity occurring in close proximity to the car using this display device. Additionally, the alarm notice can be customized in terms of sound to suit the preferences of the driver. Thus, the likelihood of a motorist missing the notification that is coming from the side of the car is extremely low (figure 2.7).

2.8.1 Proposed system

The system consists of two self-driving cars (rear car and front car) as shown in figure 2.6. Both vehicles use the VLC system as given in chapter 2 of this book to communicate. It is good to mention that in this chapter we will give a description of the collision avoidance system in brief into without going in deep analysis. We will clearly show the flowchart and the block diagram for both the transmitter and the receiver with brief description.

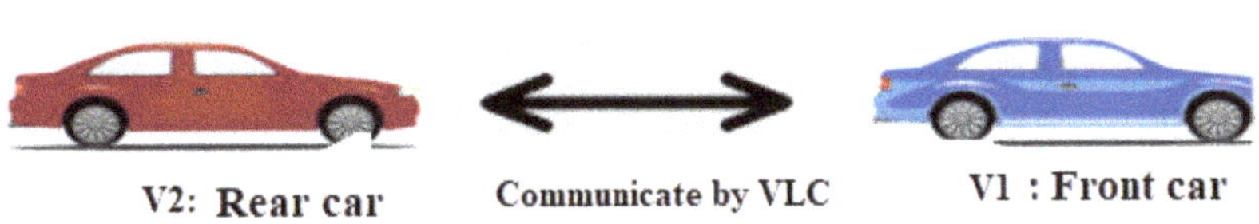

Figure 2.6. Vehicle-to-vehicle communication by VLC.

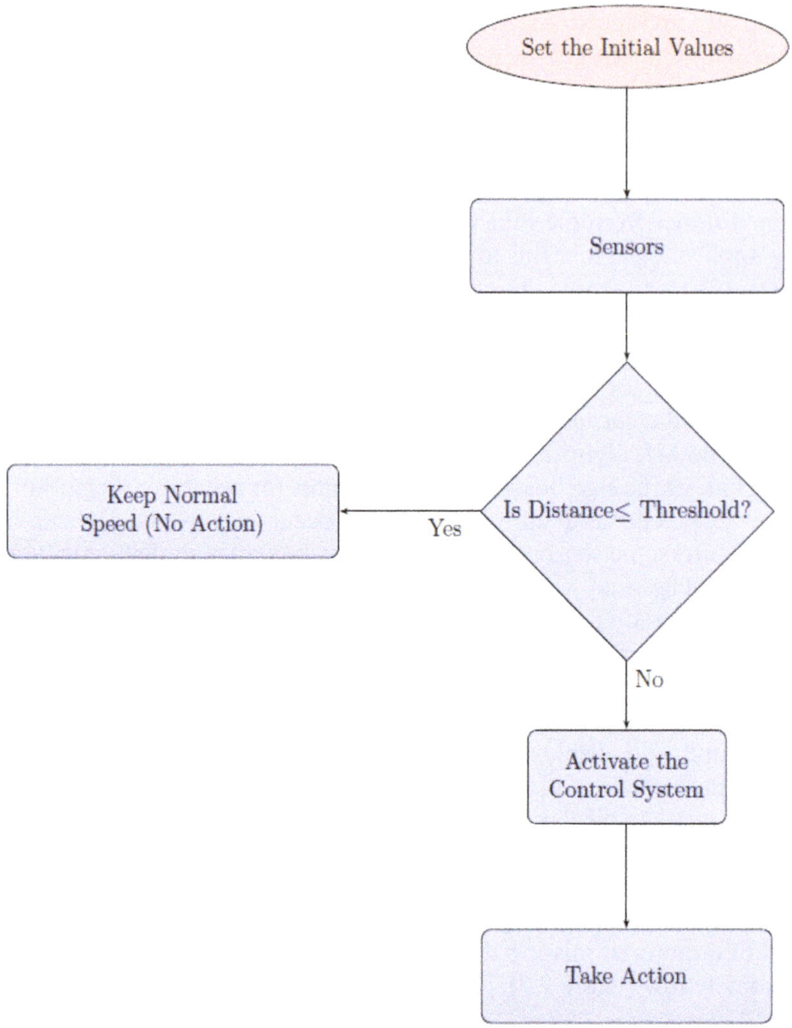

Figure 2.7. Algorithm of collision avoidance system of autonomous vehicles (AVs).

This module assists the leading car V1 to broadcast information to the following car V2 to identify and avoid collisions. It is equipped with an ultrasonic range sensor, a Li-Fi transceiver module, and a micro-controller board.

Figure 2.8 displays a block diagram of the transmitter and its real-time module, while figure 2.9 displays the transmitter flowchart. The range sensor on car V1 is activated and begins to detect the distance between the two automobiles when a rear car, V2, draws close to the leading car V1, the LCD display connected to the transmitter portion will show the message 'WARNING' if the measured distance between two cars is less than the threshold distance (d_{th}), where the threshold distance depends on many factors such as the relative speed, distance between the two vehicles, the safety distance of the brakes.

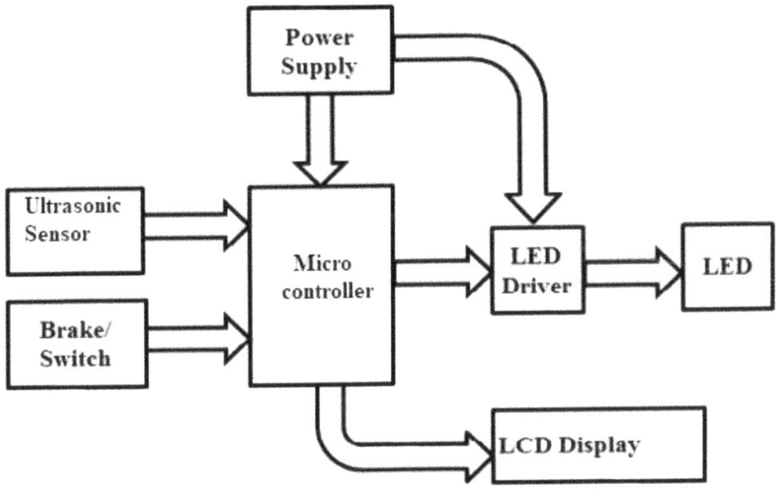

Figure 2.8. Block diagram of a transmitter.

The receiver block diagram is shown in figure 2.10. The receiver consists of a power supply, PD, amplifier, micro-controller, LCD display with a buzzer and a Li-Fi. When the ultrasonic sensor detects the distance between the two vehicles and if its less than d_{th}, the rear lights equipped with the Li-Fi technology transmit that information to vehicle V2. When the message is received, V2 will automatically begin to slow down.

In general, the ultrasonic sensor uses sonar to measure distances between objects. It offers great precision and consistent readings. Its length ranges from 2 to 400 cm, or 1 to 15 feet. Its operation is unaffected by sunlight or dark materials, such as sharp range finders (figure 2.11).

The second car's (V2) recommended speed to prevent the accident is determined by taking the measured distance (D) between the two vehicles into consideration. It receives the calculated speed of vehicle V2 through a Li-Fi transmitter.

2.8.2 Design of algorithm

The following algorithm aims to shows how to regulate the speed of car V2 in case of AVs. The algorithm is divided into two parts, namely transmitter side and receiver side.

The computed speed from the transmitter part is transferred to the receiver part based on the threshold distance, and the rear car V2 adjusts its speed accordingly.

The transmitter algorithm

Step 1: Start the execution of the command and use the range sensor to determine the distance between the two vehicles.

Step 2: Check if the distance $< d_{th}$.

Step 3: If Step 2 is true, display 'WARNING' on LCD.

Step 4: If Step2 is false, display 'SAFE ZONE!' on LCD.

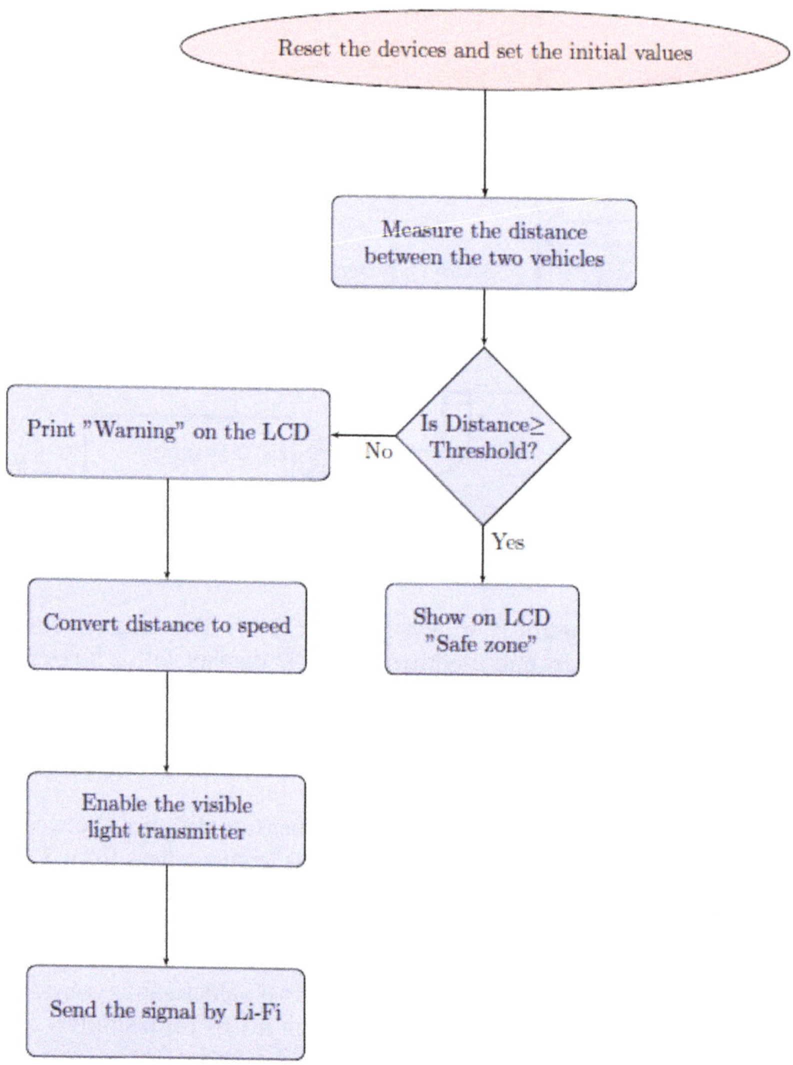

Figure 2.9. Proposed VLC collision-avoiding transmitter flowchart.

Step 5: Calculate the speed from the measured distance.
Step 6: Send the determined speed to the receiver.

The receiver algorithm

Step 1: Receive the speed from the transmitter, and display it on LCD.
Step 2: Convert speed to degree of rotation (DOR).
Step 3: Compare the current position ⩾ Layout position of the pointer.
Step 4: If the current position ⩾ Layout position ($C_{pos} > L_{pos}$), is true, rotate pointer anti-clockwise.
Step 5: If Step 4 is false, rotate pointer clockwise.

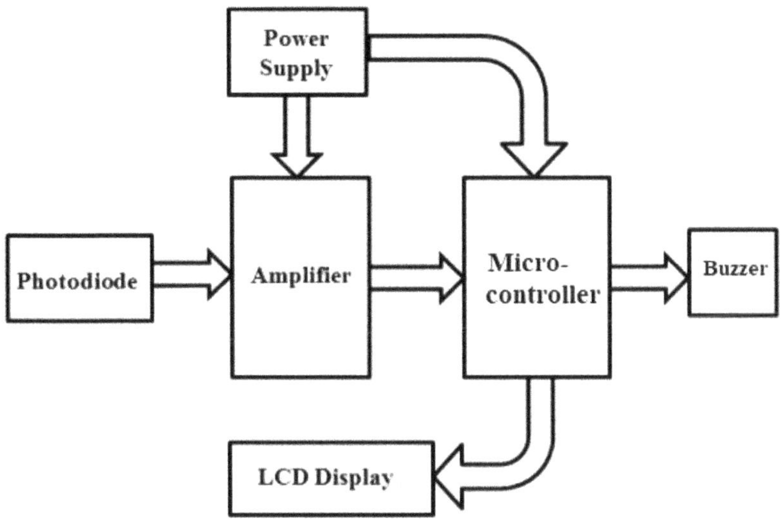

Figure 2.10. Block diagram of a receiver.

Upon receiving the transmitted speed, the vehicle V2 displays the value on its LCD. To represent the speed of car V1, the received speed is translated to DOR in the proposed system. The speedometer needle rotates counterclockwise if the current position C_{pos} of the pointer is greater than the Layout position L_{pos}, indicating that the vehicle V2 is traveling slower to prevent a collision; otherwise, it turns clockwise, indicating that the likelihood of a collision is reducing and vehicle V1 is traveling faster (table 2.6).

2.8.3 Simulation and results

MATLAB software and Proteus software are used to simulate the proposed system, as shown in figure 2.12. The simulation parameters are shown in table 2.6. In this simulation, and due to an ultrasonic range sensor's limited measuring range, we consider 1 m distance is the safety distance between two cars. LoS path and NLoS path are considered. We assume one reflector in the path. We consider the acceleration up to 100 km h^{-1} for both vehicles and they move in the same direction. We use a Philips Ultinon LED 12 985BWX2 as transmitter. Also, we use APD model (Si) Hamamatsu-S8664-1010 as a receiver. We analyze the system performance considering various parameters.

We used a function generator and an oscilloscope offered by Proteus software to measure the behavior of the transmitter and receiver outputs at a distance of 1 m after developing the final circuit for the system.

Figure 2.13 shows a simulation of how a Kalman filter has been used to track the objects. As an example when $t = 0$, the distance between the vehicles is 100 m, in this case the distance is greater than the threshold (90 m), hence the avoidance system will be deactivate, but at $t = 1$ the distance between vehicles is 80 m where the

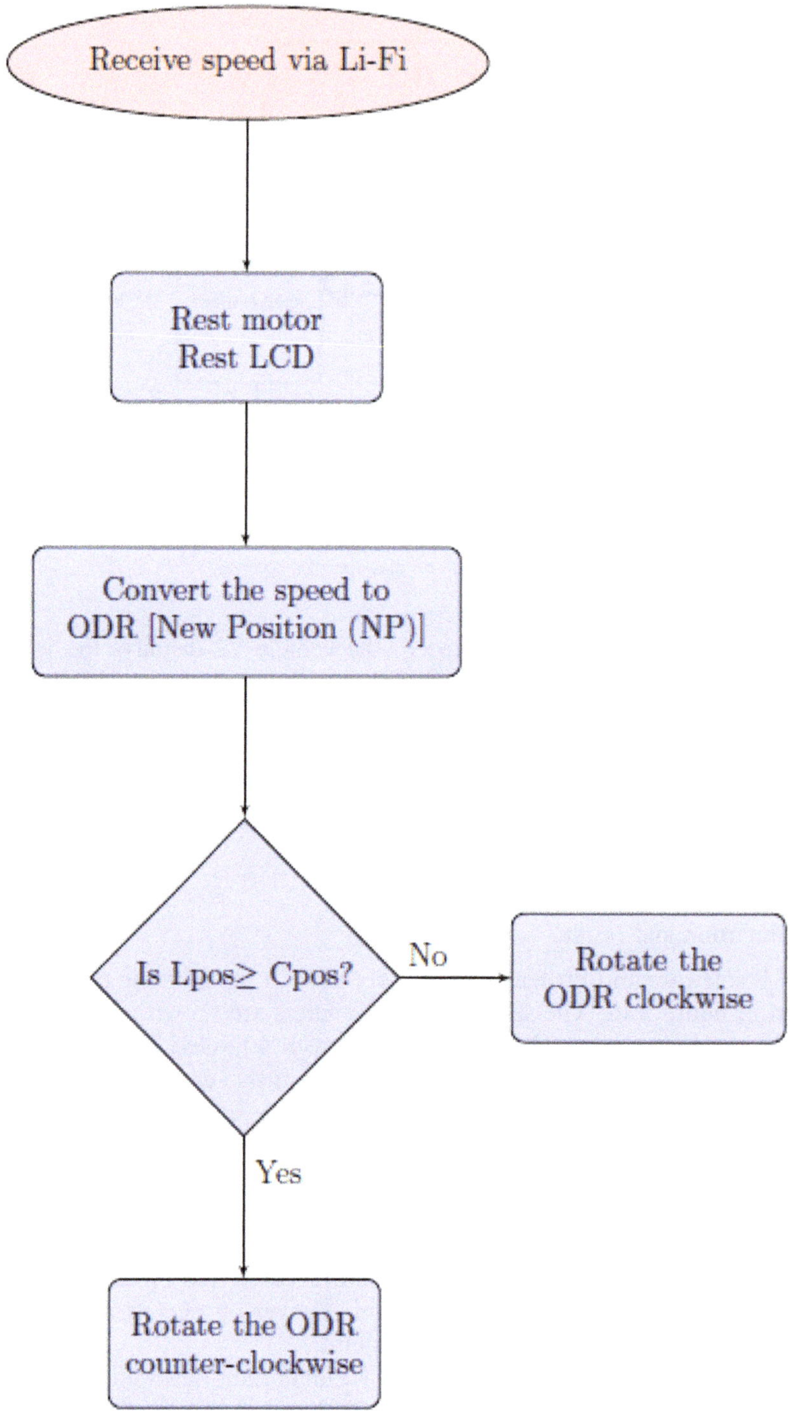

Figure 2.11. Proposed VLC collision-avoiding receiver flowchart.

Table 2.6. Simulation parameters.

Parameter	Value
Simulation time	275 s
Number of AVs	2
Vehicular range	20–120 m
Transmitted power	−10 dBm
Height of the transmitter and the receiver	20–120 cm
Angular of front and rear light	45°
Speed	20–100 km h^{-1}
Safety distance	1 m
Modulation type	OOK
Number of reflectors	1
Receiver area	4 cm^2
LED luminance intensity	1000–2500 cd
Responsivity of PD γ	0.54 A W^{-1}
Gain of the optical filter $T(\Psi)$	1
Gain of the optical concentrator g	5
Background noise	0 dBm
Sinusoidal wave	$10Vp - P$, 1 kHz
Data rate	35 Mbps

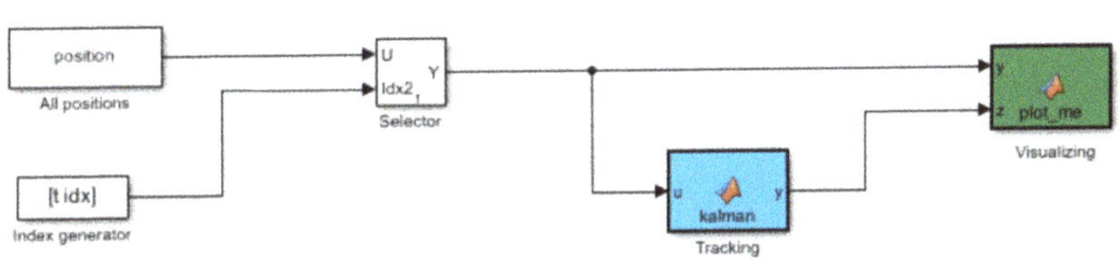

Figure 2.12. MATLAB function block to estimate the location of a moving object.

distance in this case is less than threshold distance, hence the system will be activated to avoid collision.

The link between received power and distance is considered first. Because of the high light intensity at the receiver photodiode, the received power increases as the distance between the transmitter and receiver gets closer, even when other parameters remain constant (figure 2.14).

We studied the system performance in terms of bit error rate (BER) and the distance between vehicles as shown in figure 2.15. The figure shows the inversely proportional relation between the BER and the distance.

Based on the relationship between SNR fluctuations and distance variations (figure 2.16), it can be determined that the highest SNR value of 15 dB is obtained at 20 m, and the minimum SNR value of −5 dB is obtained at 120 m. This highlights

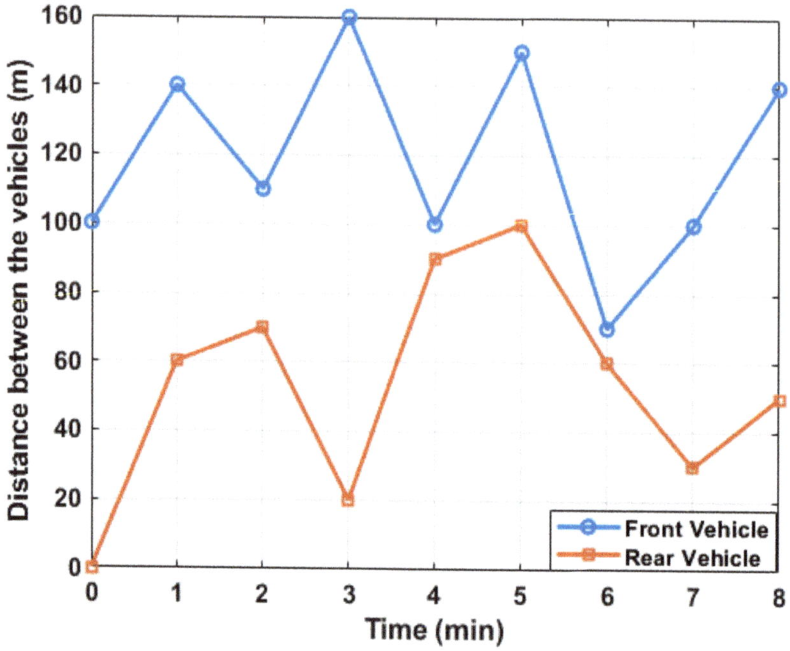

Figure 2.13. Tracking the vehicles using Kalman filter.

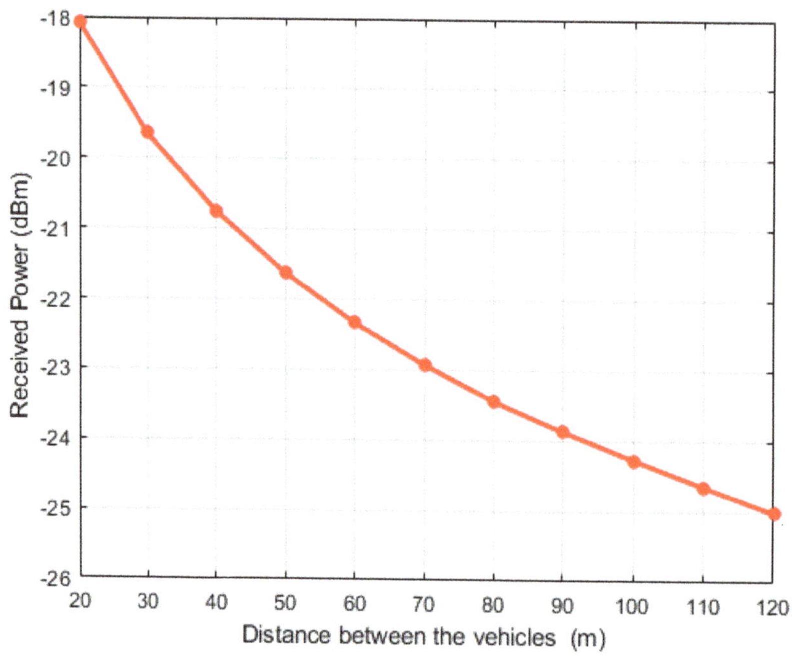

Figure 2.14. The system performance in terms of received power and the distance between vehicles.

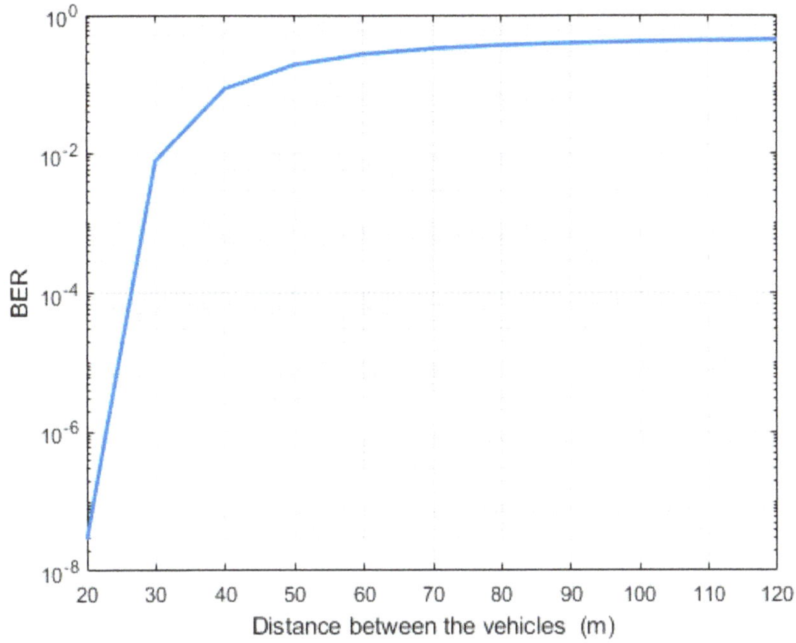

Figure 2.15. The system performance in terms of BER and the distance between vehicles.

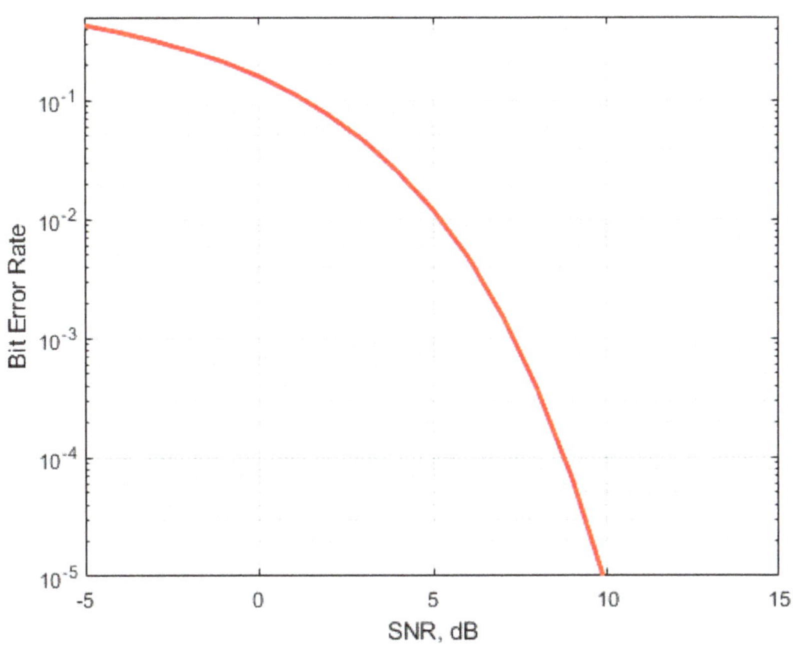

Figure 2.16. System performance in terms of BER and SNR.

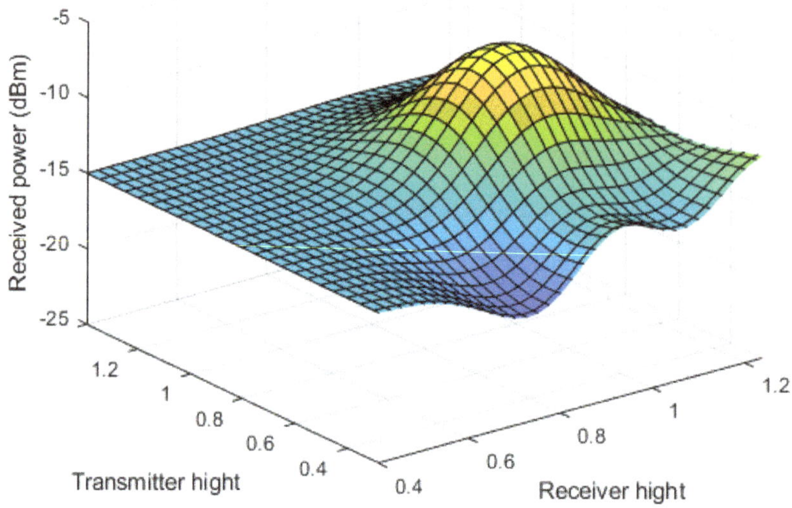

Figure 2.17. The system performance in terms of received power and the hight of both the transmitter and the receiver.

the fact that when the distance gets closer to the minimum value assumed, the signal quality gets better.

Finally, the distance relationship over BER is shown in the image. The BER value is clearly related to the distance, as can be shown. At a minimum distance of 30 m, the BER percentage approaches 0. However, at a distance of 71 m, it is the highest, measuring 1.3280×10^{-1}. The inversely proportional relationship between SNR and BER is illustrated in figure 2.16 by displaying the SNR values and matching BER percentages. For example, for SNR = 10 dB, the corresponding BER $= 1 \times 10^{-5}$, indicating that a high SNR value leads to a low BER.

Our analytical results show that utilizing the VLC technique, it is possible to establish an LoS communication up to 71 m, which is sufficient for a dependable communication link between the infrastructure and cars.

It is well known that any tail- or headlight for the vehicle should be located at a height of not more than 182.88 cm nor less than 50.8 cm.

In order to improve the system's viability and efficiency, the BER is displayed for a range of distances versus different transmitter heights (h) from the ground as well as irradiance/incidence angles. Figure 2.17 shows that the maximum received power can be achieved when the height is 0.9 m for both the transmitter and receiver.

2.9 Chapter summary

This chapter introduces the basic elements of VLC systems for vehicular applications. It describes the basics of VLC transmitters, receivers, and the channel. It compares VLC with traditional RF communication in vehicular environments and discusses the challenges of communication between LED-based VLC emitters (e.g., vehicle headlights or taillights) and PD or image sensor-based VLC receivers. The chapter also covers different types of noise and SNR under APD and PIN PDs. Finally it presents an example collision avoidance system with hardware descriptions and simulation results.

References

[1] Andonovic I, Beshr M and Hussien M 2012 The impact of sunlight on the performance of visible light communication systems over the year *Proc. SPIE—The International Society for Optical Engineering* 1387–91

[2] The relationship between the illuminance and the candela Light termenology Available online: http://www.greenbusinesslight.com (Accessed: July 10, 2024)

[3] Wang Z, Zhong W-D, Yu C, Chen J, Francois C P S and Chen W 2012 Performance of dimming control scheme in visible light communication system *Opt. Express* **20** 18861–8

[4] Walewski T, Sphicopoulos J, Ntogari T and Kamalakis G 2011 Combining illumination dimming based on pulse-width modulation with visible-light communications based on discrete multitone *IEEE/OSA J. Opt. Commun. Netw.* **3** 56–65

[5] Mesleh R, Elgala H and Haas H 2011 On the performance of different OFDM based optical wireless communication systems *IEEE/OSA J. Opt. Commun. Netw.* **3** 620–8

[6] Terra D, Kumar N, Lourenço N, Alves L N and Aguiar R L 2011 Design, development and performance analysis of DSSS-based transceiver for VLC *2011 IEEE EUROCON—Int. Conf. on Computer as a Tool* 1–4

[7] Le Minh H, O'Brien D, Faulkner G, Zeng L, Lee K, Jung D, Oh Y and Won E T 2009 100-Mb s^{-1} NRZ visible light communications using a postequalized white LED *IEEE Photon. Technol. Lett.* **21** 1063–5

[8] Hranilovic S 2005 *Wireless Optical Communication Systems* (New York: Springer)

[9] Yu S-H, Shih O, Tsai H-M, Wisitpongphan R D and Roberts R 2013 Smart automotive lighting for vehicle safety *IEEE Commun. Mag.* **12** 50–9

[10] Komine T and Nakagawa M 2004 Fundamental analysis for visible-light communication system using LED lights *IEEE Trans. Consum. Electron.* **50** 100–7

[11] Lee S, Kwon J K, Jung S-Y and Kwon Y-H 2012 Evaluation of visible light communication channel delay profiles for automotive applications *EURASIP J. Wirel. Commun. Netw.* **2012** 370

[12] Peebles J R and Peyton Z *Probability, Random Variables, and Random Signal Principles* 4th edn (New York: McGraw-Hill)

[13] Agrawal G P 2012 *Fiber-Optic Communication Systems* 4th edn (New York: Wiley)

[14] vanRooyen P and Solms F 1995 Maximum entropy investigation of the inter user interference distribution in a DS/SSMA system *Proc. 6th Int. Symp. on Personal, Indoor and Mobile Radio Communications* **vol 3**

[15] Lu Q F, Ji X S and Huang K Z 2014 Clipping distortion analysis and optimal power allocation for ACO-OFDM based visible light communication *2014 4th IEEE Int. Conf. on Information Science and Technology (ICIST)* 320–3

[16] Takai I *et al* 2013 LED and CMOS image sensor based optical wireless communication system for automotive applications *IEEE Photon. J.* **5** 6801418

[17] Photo diode parameters Available online: http://www.hamamatsu.com (Accessed: June 28, 2024)

[18] Illanko K and Fernando X 2021 Adaptive minimization of direct sunlight noise on V2V-VLC receivers *2021 IEEE 93rd Vehicular Technology Conf. (VTC2021-Spring)* 1–6

[19] Diller J and Fernando X 2022 A study of VLC between vehicles and traffic signal lights *Computer Networks, Big Data and IoT* ed A Pasumpon Pandian, X Fernando and W Haoxiang (Singapore: Springer Nature) pp 643–55

[20] Căilean A-M 2014 Study, implementation and optimization of a visible light communications system. Application to automotive field *PhD Thesis* Ştefan cel Mare University of Suceava

Chapter 3

Impacts of directional propagation

This chapter explores the application of visible light communication (VLC) in vehicular contexts, focusing on the advantages of directional light propagation that enhances secure communication between vehicles and traffic lights. It addresses the challenges of line-of-sight (LoS) requirements, obstacles, and shadowing effects that can impede signal quality. Techniques like adaptive tracking using extended Kalman filters aim to optimize signal-to-noise ratio (SNR), especially in sunlight interference scenarios. The chapter emphasizes the importance of integrating VLC within existing traffic systems to improve safety and efficiency while analyzing potential methods for broader deployment despite technical limitations.

3.1 Traffic light-to-vehicle communication scenario

The directional propagation nature of light waves in VLC significantly influences its application and effectiveness. Unlike radio waves used in traditional wireless communication, VLC utilizes light waves, which are inherently directional. This characteristic allows VLC systems to offer more secure and localized communication, as the transmission can be confined within a specific area or directed towards intended receivers without significant interference from other sources. Moreover, the directional nature enables precise targeting of communication, reducing the chances of eavesdropping or interception by unintended recipients. However, VLC also faces challenges related to LoS requirements, where obstacles or physical barriers can obstruct or attenuate the signal, potentially limiting its range and reliability in certain environments.

There are many applications in which wireless communication between vehicles and traffic lights can be very beneficial. This will enhance traffic management and safety by enabling real-time updates on traffic conditions, signal changes, and emergency notifications, thereby improving overall traffic efficiency and reducing congestion.

doi:10.1088/978-0-7503-6049-4ch3

However, this internet-to-vehicle (I2V) VLC also comes with challenges that need careful consideration. One major limitation is its requirement for direct LoS between the traffic light (transmitter) and vehicles (receiver). Obstacles such as buildings, vehicles, or adverse weather conditions like fog or heavy rain can obstruct the light signals, potentially causing communication disruptions. Furthermore, VLC is energy-efficient only when using LEDs for transmission.

The practical improvement of applications such as traffic lights through the use of VLC remains uncertain, yet the technology holds significant promise in the automotive sector due to its minimal infrastructure requirements and compatibility with existing light-based systems. Despite current challenges like lower SNR and limited sensing range, the potential for enhanced safety and traffic efficiency is substantial with large-scale integration.

Addressing the challenges mentioned could lead to a considerable increase in data exchange and cooperation among vehicles on the road. While drivers would retain the autonomy to make decisions, the ramifications and insights from those decisions could be rapidly shared across the network of motorists.

The initial goal is to compare various methods of integrating VLC systems into existing traffic signals, assessing the strengths and weaknesses of each. Research has established metrics such as SNR, range, and integration capability to evaluate these techniques. Enhancing traffic lights is crucial as it can positively impact numerous vehicles and areas, but it requires a cost-effective solution to avoid a massive infrastructure overhaul. Understanding the bottlenecks that prevent the large-scale adoption of I2V communications is vital. The current traffic management infrastructure is deeply rooted and would need significant modification to accommodate new VLC systems. While the benefits in safety could be substantial, it must be evident beforehand that such improvements are achievable with VLC. Additionally, these systems must evolve to comply with the rapidly advancing autonomy of vehicles.

A simple and effective solution currently in use is an optical collection device. This device acts as a shield that limits the acceptance angle of incoming light on the sensor. Typically, these are used alongside a lens to gather the light that is not attenuated by the shield and focus it onto the detector. The acceptance angle of light that reaches the sensor is commonly known as the field of view (FoV).

The trade-off when using such a device is that a low FoV reduces the amount of unwanted noise light but also decreases the effective range over which a moving detector remains axially aligned with the transmitter. Please refer to figure 3.1.

As depicted figure 3.1, a low sensor angle increases the $d_{\max}$ value. However, when the transmitter is close to the detector, such as when a car is stopped at a red light, a shallow sensor angle may not be sufficient to view the transmitter.

From the figure the following relationships were obtained solving for distance value range owing to the set of angle values, θ and FOV.

$$d_{\min} = \frac{h_2 - h_1}{\tan\left(\theta + \dfrac{\text{FOV}}{2}\right)} \tag{3.1}$$

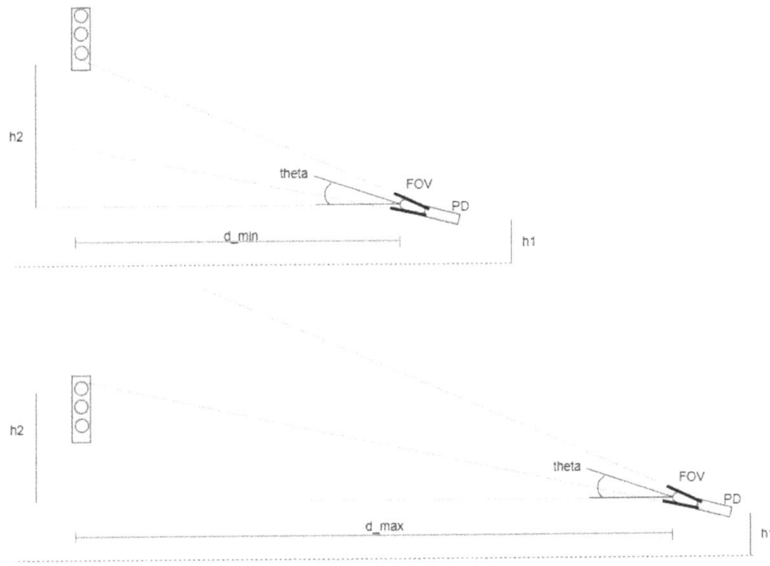

Figure 3.1. Layout of a photodiode (PD) receiver. Theta is the reception angle with respect to the horizon, FoV is the field of view, h_1 and $h2$ are the height of the receiver and transmitter, respectively.

$$d_{\text{max}} = \frac{h_2 - h_1}{\tan\left(\theta - \dfrac{\text{FOV}}{2}\right)} \tag{3.2}$$

Figures 3.2 and 3.3 show the maximum total range of the sensor, ($d_{\text{max}} - d_{\text{min}}$). Another aspect to consider is the importance of the minimum distance, denoted as d_{min}. As illustrated in figure 3.2, d_{min} represents the closest distance a vehicle can be to the traffic light (transmitter). A narrow FoV and a low sensor angle will prevent the sensor from detecting the transmitter's light if the car is too close, necessitating a greater minimum distance for signal reception. Conversely, a wider FoV allows for a shorter minimum distance by increasing the range at which the light from the transmitter can be detected. Typically, for traffic lights, the transmitter's height, h, and d_{min} are 8 m and 15 m, respectively. The accompanying plot demonstrates the minimum distances at which a sensor can detect signals, based on its mounting angle, θ, and FOV. From figure 3.4, a minimum angle of $\approx 15°$ is needed in order to view at a distance of 20 m with an FoV of 8°. However, this distance decreases to 15 m when the FOV is 20°.

As indicated, enhancing the FoV reduces the minimum distances and the overall effective range at which the sensor operates effectively. However, indiscriminately increasing the FoV is problematic. A larger FoV introduces unwanted light and additional noise. A gain equation that demonstrates the correlation between the angle, ψ, and the resultant gain, $g(\psi)$ can be written as [1],

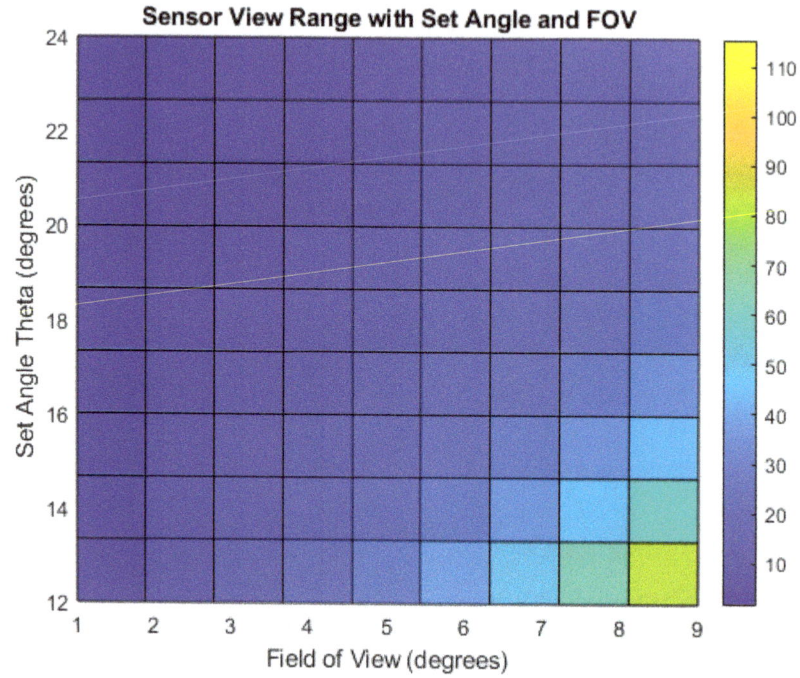

Figure 3.2. Variation of elevation angle (Theta) at the PD with FoV. Color bar values are in meters.

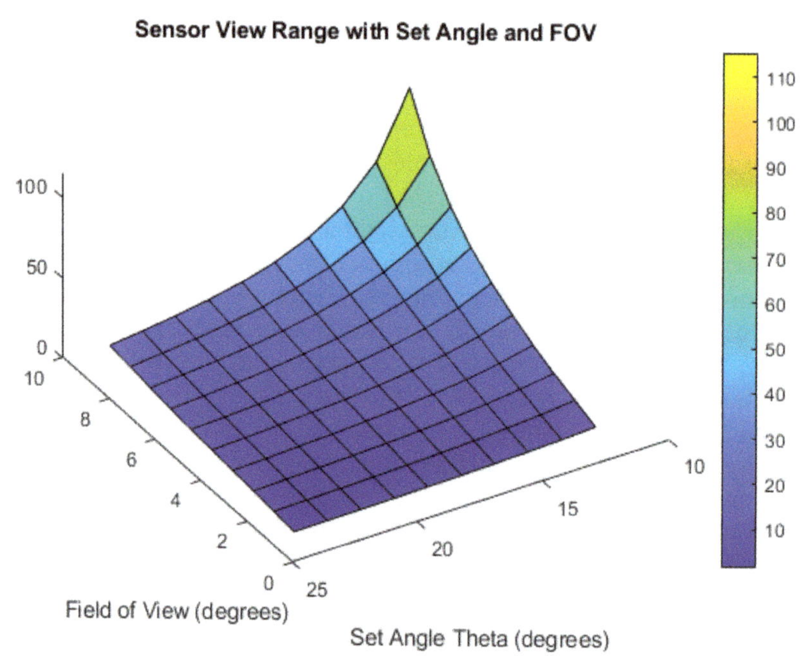

Figure 3.3. Variation of PD elevation angle (Theta) at the PD and the effective communication range with FoV. Color bar values are in meters.

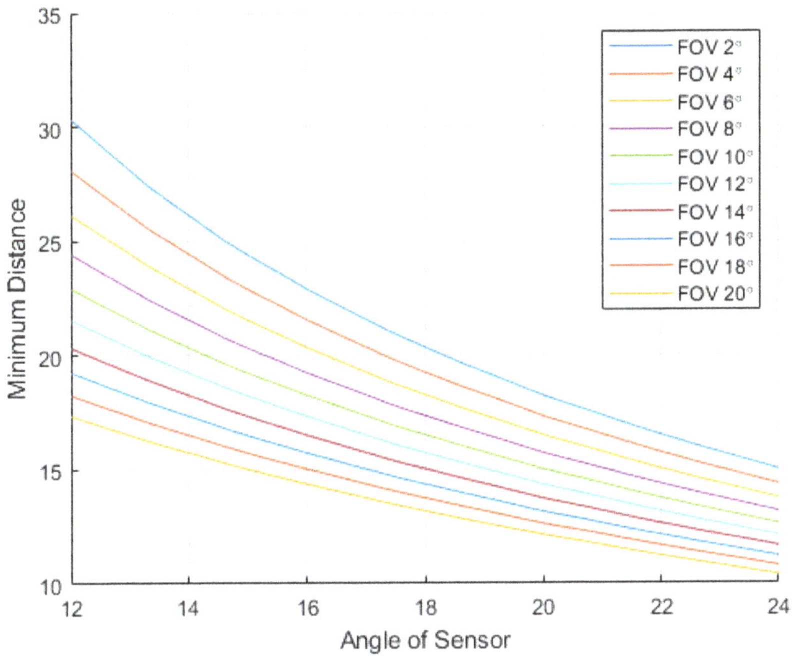

Figure 3.4. Minimum range (viewing distance) versus PD elevation angle for different FoV values.

$$g(\psi) = \frac{n^2}{\sin^2(\text{FOV})}, \qquad 0 \leqslant \psi \leqslant \text{FOV} \tag{3.3}$$

where, n represents the refractive index of the collecting lens, assumed to be 1.5.

In addition to the trade-offs for optical collection devices, there is also a trade-off for the optical transmitter, such as a traffic light. Assuming the traffic light is a Lambertian emitter, the radiant intensity, R_0, can be expressed as a function of the viewing angle, ϕ, relative to the emission axis, and the beam profile, m, as shown below:

$$R_0(\phi) = \left(\frac{m+1}{2\pi}\right)\cos^m \phi \tag{3.4}$$

According to equation (3.4), a high beam parameter, m, ensures substantial radiant power when the viewing angle is near zero. However, with increasing misalignment, there is a significant drop in power. A low viewing angle also reduces the effective minimum range of the PD, affecting usability. This issue can be mitigated by enlarging the FoV of the receiver, but this results in decreased signal gain.

Using the equation and plotting for various m parameters and θ, the received intensity is illustrated in figures 3.6 and 3.5.

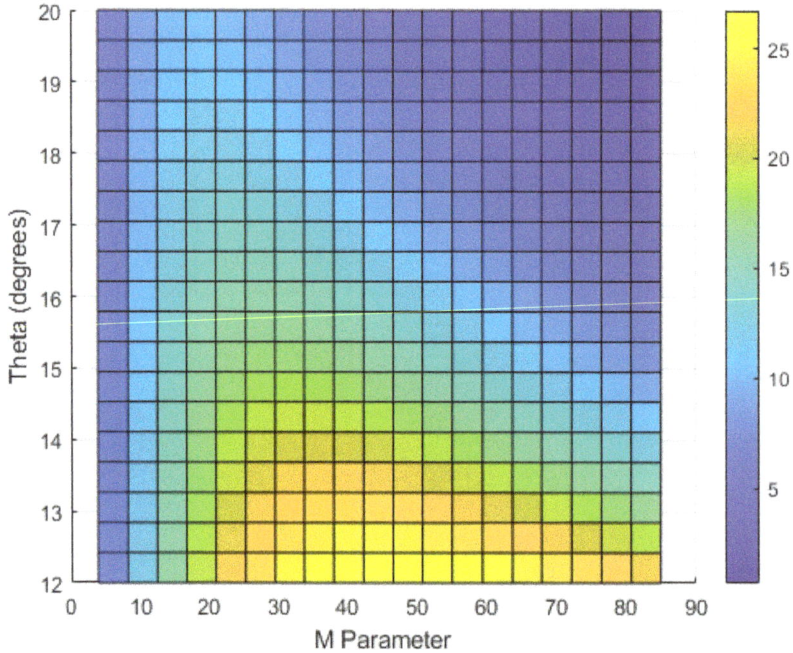

Figure 3.5. Receiver elevation angle versus M parameter.

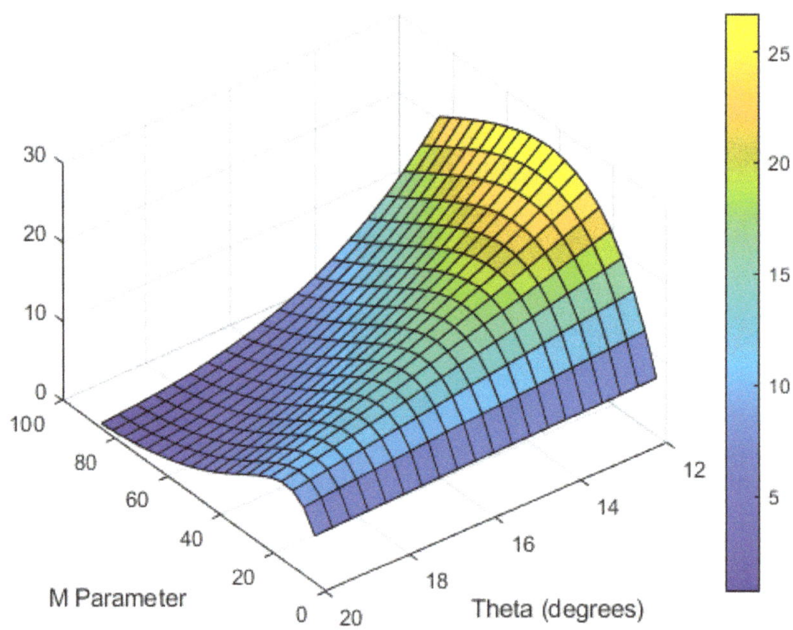

Figure 3.6. Variation of light intensity with radiation profile parameter (M) and viewing angle (Theta). $M = 1$ is a Lambertian source.

It can be observed that the minimum alignment angle of the PD should be around 16° and the optimal *m* parameter is approximately 35 to receive a reasonable amount of light.

3.1.1 Novel potential improvements

This presents a potential low-cost benefit for the transmitter by segmenting the output emission angle. If the transmitter's light were divided into several bands, each with a higher *m* parameter and a distinct emission angle, the radiant intensity could be enhanced across a broader region (see figure 3.7).

The lower area at the transmitter's edges could be angled downward with a higher beam parameter, while the central bands could maintain a low beam divergence with a shallow angle for distant receivers. This approach could mitigate the trade-off between low FoV/high gain and an extended sensor range. Each transmitter segment could have a specific attack angle to target a region with a high beam parameter spatially.

Additionally, implementing an array of detectors on the vehicle could increase the likelihood of axial alignment between the transmitter and receiver, while a narrow FoV could be used to maximize signal gain (figure 3.8).

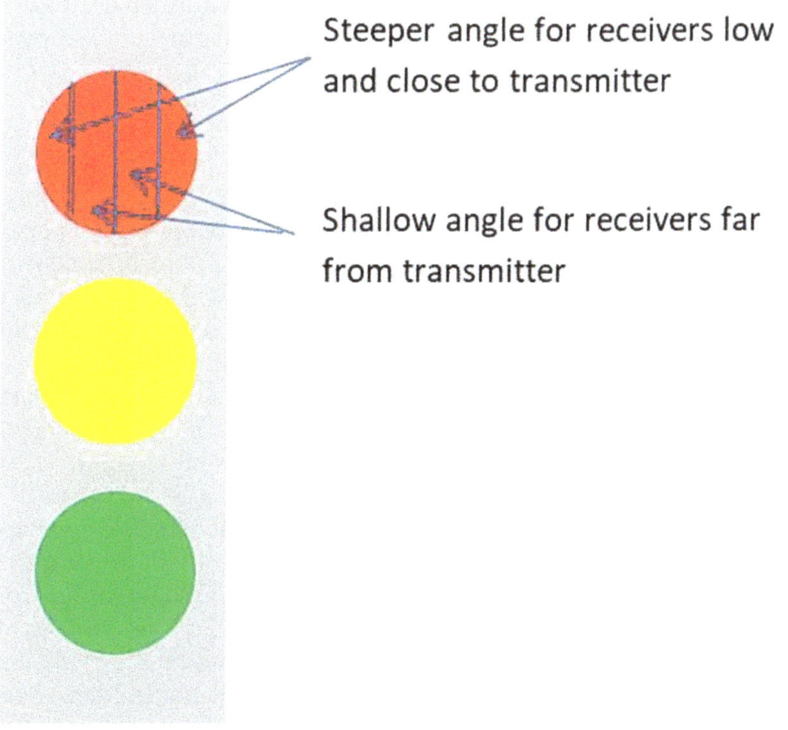

Figure 3.7. Traffic light arrangement.

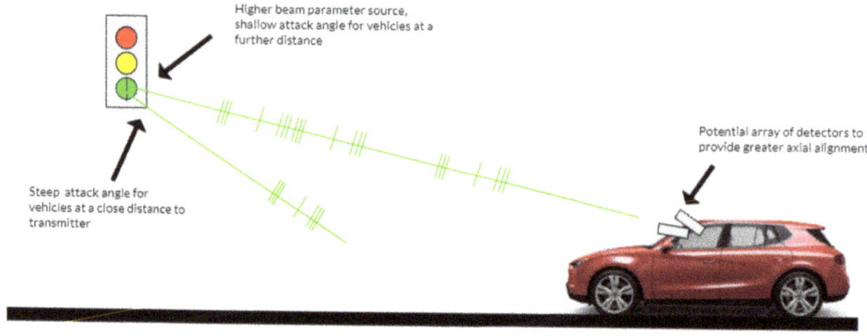

Figure 3.8. Example of array transmitter and receiver for maximized beam parameter and gain at varying vehicle approach distances.

While this suggests increased retrofit costs for widespread integration, if such changes could be implemented inexpensively, they could further enhance this technology.

From the above analysis, we can observe that technical issues such as noise and range currently prevent the large-scale integration of these types of sensors. Nevertheless, recent advancements have enhanced performance and maintained high potential for broad integration. This trend appears to benefit the widespread application of this technology and enhance safety for all.

Future technologies may merge with efforts by original equipment manufacturers and automakers to significantly boost road safety and further decrease fatalities. With ongoing advancements, large-scale integration seems inevitable if performance becomes practical and costs remain manageable.

3.2 Shadowing effects

Shadowing is the other major impairment due to directional propagation properties of light. Note that theoretically when we visually perceive an object, it is the rays of light reflected or scattered off that object that your eyes absorb. Shadow is a dark (real image) area where light from a light source is blocked by an opaque object. It occupies all of the three-dimensional volume behind an object with light in front of it. The cross-section of a shadow is a two-dimensional silhouette, or a reverse projection of the object blocking the light. An object is almost never in simply light and shade. Rather, it is usually in an environment in which light is bouncing around in several directions (see figure 3.9). For this reason, it is important for beginners to understand the nature of shadows and light.

Due to the nature of the visible light spectrum, shadowing is a significant issue, especially over short distances, and small FoVs. This can seriously impair communications in both quasi-stationary indoor and fast varying outdoor environments.

However, by employing transmitters with a wide FoV as well as effectively utilizing the diffraction phenomenon of the light, the issues caused by shadowing can be alleviated. Note, it is much better to implement diffuse transmitters which radiate the light over a wide solid angle, which does not have as severe pointing and shadowing problems when compared to point-to-point links. The diffused

Figure 3.9. Forms of shadow and light.

transmitter does not need to be aimed directly at the receiver since the radiated optical waves reflect from multiple surfaces. Nevertheless diffused light transmission has a number of other issues such as low received power and poor SNR.

Spatial diversity is commonly implemented to mitigate the effect of shadowing, where multiple transmitters and receivers are deployed around the shadowed area [6]. However, having such a setup is not always guaranteed and sustainable, henceforth it is necessary to study the effect of shadowing without having spatial diversity. This is necessary to effectively design a VLC system, along with obtaining the parameters that control its performance determined by the signal error probability under various shadowing conditions.

3.2.1 Bimodal Gaussian reception

If the receiver is fully blocked by an opaque object, then no light from the transmitter will be received and communication is not possible. However, in most practical scenarios, some amount of light will be received by the PD. As shown in figures 3.9 and 3.10, two (or more) clusters of light reception happen due to diffraction. The diffraction phenomenon helps to overcome the impact of shadowing by receiving some amount of light. When there are predominantly two light clusters, it is referred to as *bimodal shadowing*. This description is applicable in both indoor and outdoor environments [4].

Furthermore, the impacts of shadowing can be studied under three different scenarios:

- **No shadowing:** The light is directly received by the PD under LoS communication. Here the spectral density of the received signal will have two peaks (simple bimodal) corresponding to two transmit power levels (see figure 3.11).

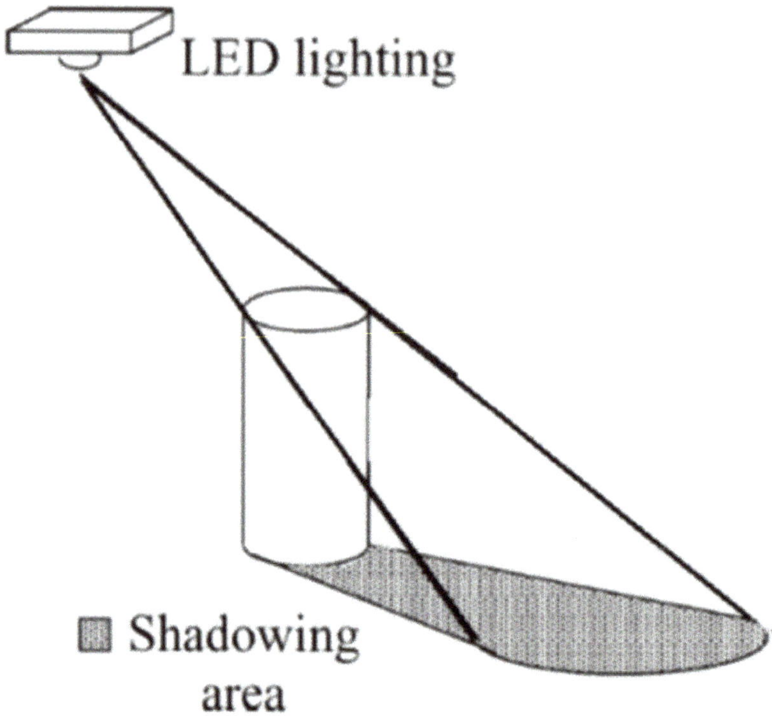

Figure 3.10. Bimodal scenario: light is received in two clusters due to partial shadowing.

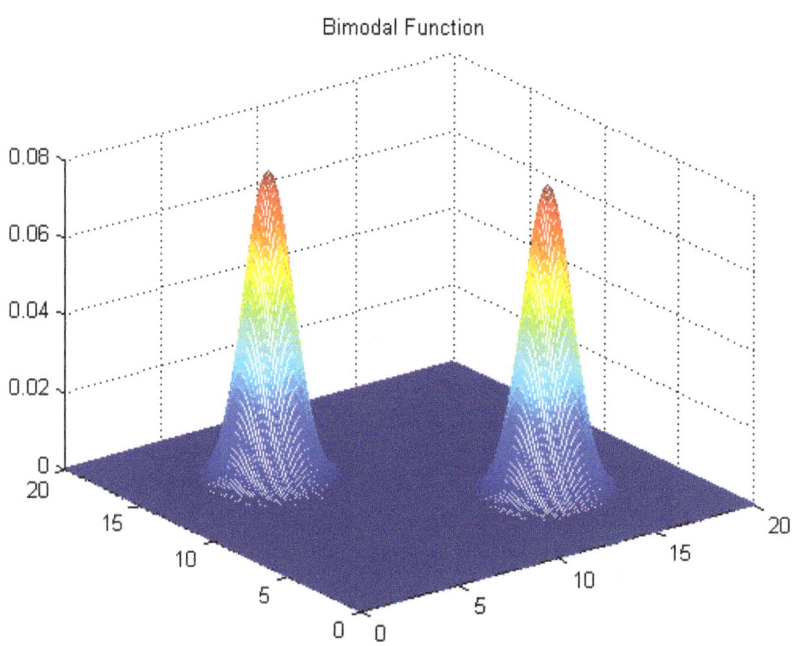

Figure 3.11. A general bimodal function with equal variance and equal amplitude. Copyright 2016, IEEE. Reprinted, with permission, from [4].

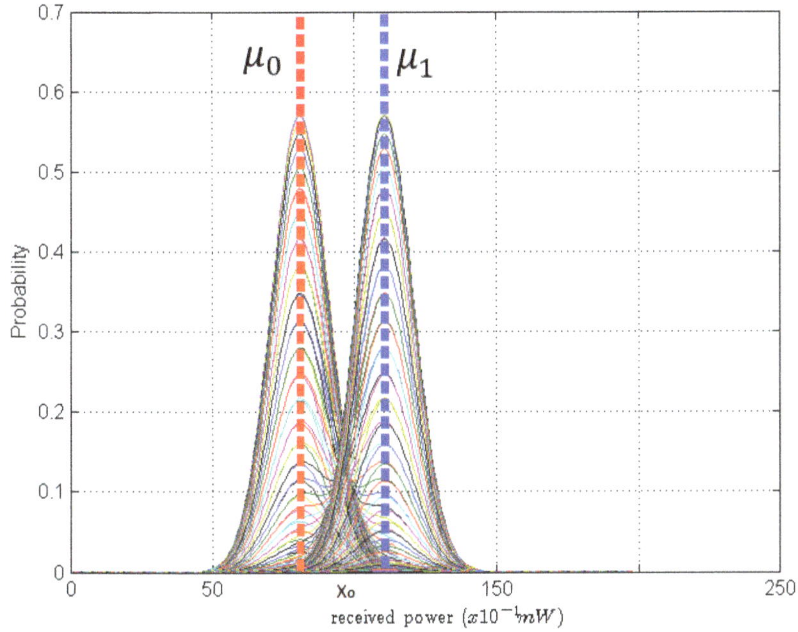

Figure 3.12. Probability density function (PDF) of the received signal without shadowing. There are two peaks, each correspond to bit '1' and bit '0'. Both distributions have equal variance. Copyright 2016, IEEE. Reprinted, with permission, from [4].

- **Partial shadowing:** Some light is received. There will be two clusters of light reception. The time difference between them is less than a bit duration. Here each transmit power level will cause multiple peaks at the received power spectral density (PSD), that are closely spaced (see figure 3.13).
- **Severe shadowing:** Blocking object is fairly large. There will be two clusters of light reception. Since time difference between them is larger than a bit duration, the received PSD will have multiple peaks crossing each other (see figure 3.14).

In this section, the effect of the shadowing on a VLC system will be discussed. Moreover, the shadowing effect for visible light will be modeled by a bimodal distribution and the probability of error for no-shadowing, moderate-shadowing and severe-shadowing conditions derived.

3.2.2 Channel impulse response with shadowing

In a typical VLC system, the channel impulse response for a bimodal obstruction is given by,

$$h_1(t) = \frac{\left(\frac{l+1}{2\pi}\right) A_r \cos^{l}(\phi) \cos(\theta)}{{R_1}^2} \delta(t - \frac{R_1}{c}) \tag{3.5}$$

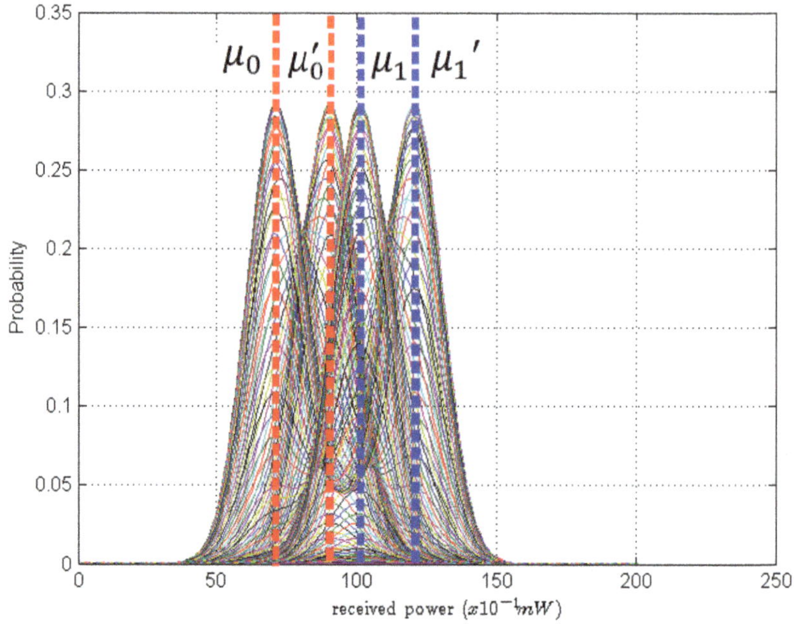

Figure 3.13. PDF of the received signal with moderate shadowing. There are four peaks, bit '1' corresponds to two peaks and bit '0' also corresponds to two peaks due to shadowing. Copyright 2016, IEEE. Reprinted, with permission, from [4].

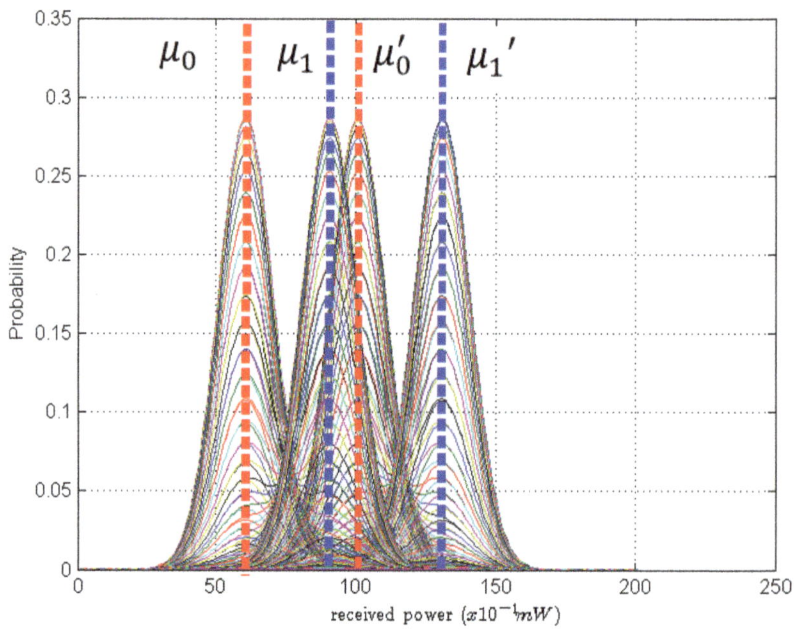

Figure 3.14. PDF of the received signal with severe shadowing. There are four peaks, bit '1' corresponds to two peaks and bit '0' also corresponds to two peaks due to shadowing. However, the peaks cross each other significantly increasing the error rate. Copyright 2016, IEEE. Reprinted, with permission, from [4].

$$h_2(t) = \frac{\left(\frac{l+1}{2\pi}\right) A_r \cos^{l}(\phi) \cos(\theta)}{{R_2}^2} \delta(t - \frac{R_2}{c}) \tag{3.6}$$

where, l denotes the order of the source, R_1 and R_2 are the total distance from the lighting source to the receiver, taking into account the light diffraction around the obstructing object from the first edge and the second edge of the obstructing object, respectively, ϕ is the emission angle, θ is the incident angle, and c is the speed of the light. Note, the order l is based on the equation,

$$B(\theta) = B_o \cos^l(\theta)$$

where, the radiance at angle θ given by $B(\theta)$, depends on θ as well as l. For a Lambertian light source $l = 1$. Higher l will result in higher directivity. Half power angle is 60 degrees for a Lambertian source indicating wide-angle radiation.

The received signal energy E_r is,

$$E_r = \int_0^T |x(t) \otimes (h_1(t) + h_2(t))|^2 \, dt. \tag{3.7}$$

where, $x(t)$ denotes the transmitted signal, T denotes the bit duration, and $\otimes$ sign refers to convolution.

3.2.3 Power spectral density of the received signal

In a standard VLC system without shadowing, it can be assumed that the transmitted signal encounters additive white Gaussian noise (AWGN), hence the distribution at the receiver remains Gaussian. In short-distance LoS links, a multipath dispersion is insignificant. Hence, LoS links channel are often only modeled with a linear attenuation and delay [2]. Optical LoS links are considered as non-frequency selective where the path loss depends on the square of the distance between the transmitter and the receiver [4].

Since shadowing plays a significant role in VLC, the final distribution will result in a multi-mode distribution depending on the severity of the shadowing effect as well as the complexity of the obstructing object (i.e. non-uniform objects with complex cavities cause secondary interference effects thereby varying the channel gain) but a bimodal distribution will be considered here for simplicity.

As apparent in figure 3.10 (assuming the light only passes over two edges), a bimodal power due to shadowing distribution can be considered. With a bimodal approach, and assuming the PD receiver holds a large FoV, the optical power will be received from two distinct edges passing the object, each with different polarization. Therefore, depending on the severity/complexity of the object surface, each polarizing path passing the object will result in a local maximum.

Two types of bimodal distribution are mentioned in this section.

1. Bimodal distribution of the received signal due to two transmit power levels corresponding data bits '1' and '0'.

2. Multi- or bimodal distribution of the received due to multiple clusters of light received due to shadowing.

There should be no confusion between these two.

3.2.3.1 PSD without shadowing

Under LoS conditions, the received signal will have two peaks with Gaussian distribution (bimodal). Each peak will be representing bit '1' and bit '0', respectively. The separation between the peaks depends on the power difference under the 'on' and 'off' conditions of the transmitter.

Equation for the distance between the two peaks of the resulting signal $S(t)$ at time t is given by [7],

$$S(t) = \frac{\mu_1(t) - \mu_0(t)}{2(\sigma_0(t) + \sigma_1(t))} \tag{3.8}$$

where μ_0 and σ_0 are the mean and the standard deviation of the first Gaussian distribution (bit '0'), while μ_1 and σ_1 are the mean and the standard deviation of the second Gaussian distributions (bit '1'), and $\mu_1 > \mu_0$.

As shown in figure 3.11, the bimodal distribution has the following PDF [3]

$$\begin{aligned} f(z, \mu_0, \sigma_0, \mu_1, \sigma_1) = \frac{1}{2\sqrt{2\,\pi\,(\sigma_0)^2}} \exp\left(\frac{-(z-\mu_0)^2}{2\,(\sigma_0)^2}\right) \\ + \frac{1}{2\sqrt{2\,\pi\,(\sigma_1)^2}} \exp\left(\frac{-(z-\mu_1)^2}{2\,(\sigma_1)^2}\right) \end{aligned} \tag{3.9}$$

3.2.4 Error probabilities with and without shadowing

In this study, the error probability derivation is with shadowing effect under a bimodal distribution case assuming on–off keying (OOK) modulation is employed, also, we assume that LEDs have a Lambertian radiation pattern [5]. This analysis can, however, be extended to a multi-modal case.

The aim is to derive the probability of error, P_e of the received signal for the three cases of shadowing: no-shadowing, moderate and severe shadowing, considering figures 3.12, 3.13, and 3.14 to illustrate their effect.

In this study, a closed form expression is derived for the probability of error for each case of shadowing. The derivation is given for the first case, while, the other cases can be derived using a similar procedure.

It is important to note that in the case of no shadowing, the transmitted light will be received without being significantly deteriorated by the shadowing object, thus, the received signal will be modeled as a Gaussian distribution for both cases of transmitting 1 and 0 represented by s_1 and s_0, with average values μ_0, μ_1, respectively, as illustrated in figures 3.12.

In the case of having moderate or severe shadowing, the light will be more affected by the shadowing object, hence the signal will be received from both sides which will result in a shadow-caused-bimodal distribution. This is clearly depicted in figure 3.13, and figure 3.14, where two Gaussian distributions are created for each received bit, thereby resulting in a bimodal distribution.

The average expected received values can be represented by four vertical lines, with two possible average values for the case of receiving s_1, represented with average values μ_1, μ_1' and two possible average values for the case of receiving s_0 represented by μ_0, μ_0'.

However, considering the fact that the shadowing is moderate, the two average values of receiving s_0 are close to each other which also applies for the case of receiving s_1. Therefore, there are two decision regions for the probability of error calculation.

With severe shadowing, the average values of the two modes are shifted significantly from each other. In this case, the shift is large enough to approximate the error by representing each mode as a normal distribution. This can be considered when the separation, $S(t) > 1$ between modes, then the distribution becomes two normal distributions, as shown in figure 3.14 [4].

Moreover, under the severe shadowing case, the Gaussian distributions of s_0 and s_1 become close to each other, which significantly affects the probability of error, which will be verified in the simulation section.

Furthermore, due to the obstruction effects, the width of each Gaussian distribution is halved when compared to the case of no shadowing due to the transmitted signal power being distributed between the two paths around the obstructing object.

In the following, the derivation of the probability of error P_e can be given as,

$$P_e = P(0)\, P(e\backslash 0) + P(1)\, P(e\backslash 1) \tag{3.10}$$

where $P(0)$ and $P(1)$ are the probability of transmitting '0' and '1', respectively, $P(e\backslash 0)$ and $P(e\backslash 1)$ are the conditional probability for receiving '1' while '0' was transmitted and receiving '0' while '1' was transmitted, respectively, which is depicted by the shadowing in figure 3.12.

Due to symmetry, one can conclude that $P(e\backslash 0) = P(e\backslash 1)$.

As depicted in figure 3.12, the two Gaussian distributions have average values μ_0, and μ_1 representing the bits 0 and 1,respectively,

Define the decision region as [4]

$$x_0 = \frac{\mu_0 + \mu_1}{2} \tag{3.11}$$

where, x_0 is the mid-point between the two means, as shown in figure 3.12.

Then $P(e/0)$ is defined as,

$$P(e\backslash 0) = \int_{xo}^{\infty} \frac{1}{\sqrt{2\,\pi\,\sigma^2}} \exp\left(\frac{-(x-\mu_0)^2}{2\,\sigma^2}\right) \mathrm{d}x \tag{3.12}$$

Also, $P(e\backslash 1)$ is defined as,

$$P(e\backslash 1) = \int_{-\infty}^{x_0} \frac{1}{\sqrt{2\,\pi\,\sigma^2}} \exp\left(\frac{-(x-\mu_1)^2}{2\,\sigma^2}\right) \mathrm{d}x \tag{3.13}$$

Recall the Q function as,

$$Q_x = \int_{x}^{\infty} \frac{1}{\sqrt{2\,\pi}} \exp\left(\frac{-y^2}{2}\right) \mathrm{d}y \tag{3.14}$$

let $y = \frac{x-\mu_0}{\sigma}$, then, $dy = \frac{dx}{\sigma}$,

By Changing the integration limits, from (3.12), $P(e/0)$ can be given as,

$$P(e\backslash 0) = \int_{\left(\frac{x_0-\mu_0}{\sigma}\right)}^{\infty} \frac{1}{\sqrt{2\,\pi}} \exp\left(\frac{-y^2}{2}\right) \mathrm{d}y \tag{3.15}$$

then,

$$P(e\backslash 0) = Q\left(\frac{x_0-\mu_0}{\sigma}\right) = Q\left(\frac{\frac{\mu_0}{2}+\frac{\mu_1}{2}-\mu_0}{\sigma}\right) \tag{3.16}$$

Equation (3.16) can be simplified to,

$$P(e\backslash 0) = Q\left(\frac{\mu_1-\mu_0}{2\,\sigma}\right) \tag{3.17}$$

The $P(e\backslash 1)$ can be found by following same steps, where $P(e/1) = Q\left(\frac{\mu_1-\mu_0}{2\,\sigma}\right)$. Recall (3.10), the probability of error for no shadowing will be given by

$$P_e = Q\left(\frac{\mu_1-\mu_0}{2\,\sigma}\right) \tag{3.18}$$

To find the probability of error for moderate and severe shadowing when the intensity modulation is employed, the probability of error will be based on four average values due to the bimodal distribution, as shown in figures 3.13 and 3.14. Then by applying the same procedure as shown above, the probability of error for moderate and severe shadowing can be given by

$$P_e = 2\left(Q\left(\frac{\mu_1-\mu_0}{4\,\sigma}\right) + Q\left(\frac{\mu_1'-\mu_0}{4\,\sigma}\right)\right) \tag{3.19}$$

$$P_e = 4\left(Q\left(\frac{\mu_1'-\mu_0'}{4\,\sigma}\right) + Q\left(\frac{\mu_0-\mu_1'}{4\,\sigma}\right) + Q\left(\frac{\mu_1-\mu_0}{4\,\sigma}\right)\right) \tag{3.20}$$

Notice that in the case of moderate shadowing, there will still be two decision regions for detecting s_1 and s_0, while in the case of severe shadowing, the error probability is further increased as shown before due to the crossing of decision regions for detecting s_1 or s_0 which results in four decision regions.

Matlab software was used to execute the simulation of the three cases of shadowing. A Gaussian channel was considered and the symbols s_1 or s_0 have equivalent variances but different means. Additionally, we are assuming intensity modulation is employed in the system where a stationary object with two main edges is obstructing the LoS signal, hence each modulation level will exhibit two modes. The error significantly increases when shadowing is in effect [4].

Figures 3.12, 3.13, and 3.14 illustrate the power density functions of the receive signal with and without shadowing. Here, blue lines represent modulation levels s_1 and red lines represent modulation level s_0, respectively.

Figure 3.12 illustrates a clear distinction between each modulated intensity level without exhibiting dual modes since there is no shadowing.

Figures 3.13 clearly shows the bimodal shadowing effect as there are two modes for each modulated intensity level.

In figure 3.14, severe shadowing causes bimodal reception and there is an overlap between the decision region of bits 0 and 1. This greatly increases the BER.

Also, we simulate the probability of error for different cases of shadowing by varying the received signal energies as well as its variances for the bimodal distribution. These are shown in figure 3.15 that compares the error probabilities

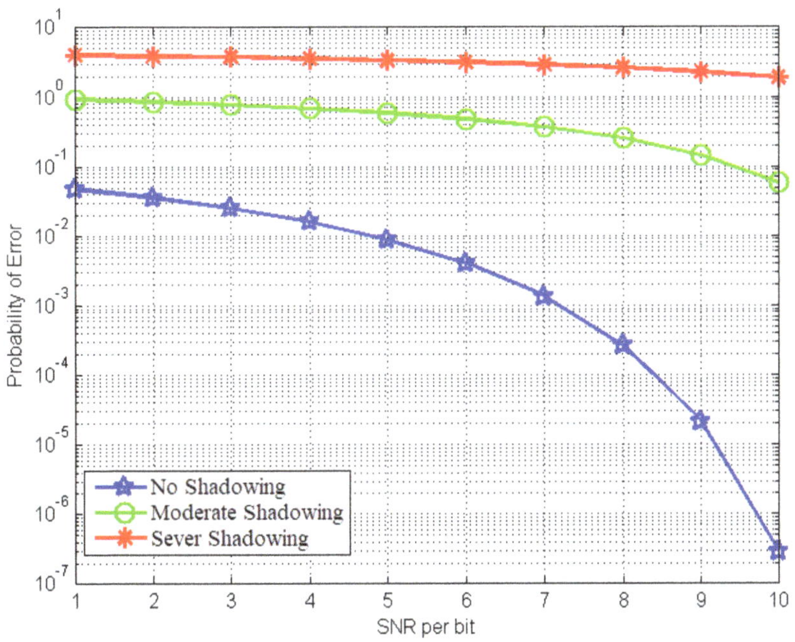

Figure 3.15. Error probability with different shadowing effects. Copyright 2016, IEEE. Reprinted, with permission, from [4].

with different shadowing effects. Clearly, the severe shadowing has the worst error probability. It is also clear that for all shadowing cases, when the received signal power is increased, as a result of increasing the LED transmitted power, the probability of error decreases.

Impacts of shadowing summary

In this section impacts of shadowing is studied under simple OOK modulation considering only two significant clusters of energy due to shadowing. As is shown in OOK, a simple form of amplitude modulation used in VLC, shadowing can cause severe performance degradation as the binary data transmission relies on distinct light intensity levels, and any obstruction can lead to data errors or loss. Pulse amplitude modulation, which uses multiple light intensity levels, might severely suffer under shadowing conditions

With higher-order modulation schemes like orthogonal frequency division multiplexing (OFDM), shadowing can still impact performance, but these schemes can better manage interference and noise, potentially allowing for more robust communication. OFDM, known for its robustness to multipath effects and better spectral efficiency, can mitigate some effects of shadowing through advanced error correction and signal processing techniques, though severe shadowing can still cause noticeable performance drops. In essence, while shadowing adversely affects VLC under all modulation schemes, the extent of impact and the system's ability to cope with it depend on the complexity and robustness of the chosen modulation technique.

3.3 Chapter summary

In this chapter, impacts due to the directional propagation aspects of light on VLC systems are considered under three different perspectives:

1. Limitations of I2V LoS communication with different heights of transmitter and receiver.
2. Vehicle-to-vehicle communication while minimizing sunlight using adaptive tracking.
3. Impacts of shadowing effects.

It can be learned that although the highly directional nature of light enhances data security and minimizes interference, this also means that VLC is susceptible to LoS obstructions, which can disrupt communication if objects block the path. Moreover, directional propagation requires precise alignment between the transmitter and receiver, making mobility and dynamic environments challenging. Overall, while directional light propagation enhances performance and security, it demands careful system design and deployment.

References

[1] Arnon S 2015 *Visible Light Communication* (Cambridge: Cambridge University Press)

[2] Carruthers J B and Kahn J M 1997 Modeling of nondirected wireless infrared channels *IEEE Trans. Commun.* **45** 1260–8

[3] Peacock N H B, Forbes C and Evans M 2005 *Statistical Distributions* 2nd edn (New York: Wiley)

[4] Farahneh H, Mekhiel C, Khalifeh A, Farjow W and Fernando X 2016 Shadowing effects on visible light communication channels *2016 IEEE Canadian Conf. on Electrical and Computer Engineering (CCECE)* 1–5

[5] Gfeller F R and Bapst U 1979 Wireless in-house data communication via diffuse infrared radiation *Proc. IEEE* **67** 1474–86

[6] Zeng L, O'Brien D C, Minh H L, Faulkner G E, Lee K, Jung D, Oh Y and Won E T 2009 High data rate multiple input multiple output (MIMO) optical wireless communications using white LED lighting *Sel. Areas Commun.* **27** 1654–62

[7] Zhang C, Mapes B E and Soden B J 2003 Bimodality in tropical water vapour *Q. J. R. Meteorol. Soc.* **129** 2847–66

Chapter 4

Channel modeling for V2V-VLC system

The visible light communication (VLC) channel, particularly in outdoor vehicular settings, is highly complex due to the dynamic nature of the environment and numerous reflections. These factors lead to losses and turbulence-induced fluctuations in the amplitude of the received optical signal. It is essential to understand key performance metrics like *coherent time* and *coherent bandwidth* to design a vehicle-to-everything (V2X)-VLC system effectively. While many resources accurately describe these metrics in RF environments, they are less prevalent for VLC environments [36].

This chapter presents a novel vehicle-to-vehicle (V2V)-VLC system in an environment with rich scattering, taking into account both line-of-sight (LoS) and non-LoS (NLoS) paths. The typical scenario involves using a pair of headlights as transmitters on one vehicle and a pair of taillights as receivers on another, akin to a 2×2 multiple-input–multiple-output (MIMO) communication link. We then formulate closed-form expressions for the channel impulse response (CIR) and assess metrics such as channel coherent time and coherent bandwidth across various scenarios, including when vehicles are moving in the same or opposite directions.

4.1 Channel modeling for VLC

Accurate VLC channel models are necessary for better system design. Using VLC in outdoor applications has a few challenges such as signal loss during severe weather conditions, sunlight, and ambient light. Weather conditions such as heavy fog, rain or snow could decrease the communication range. These degrading effects can be minimized by using highly sensitive receivers. Another strong concern is direct sunlight or strong ambient light, which could saturate the receiver [32]. Characterization of a communication channel is performed by its channel impulse response, which is then used to study, analyze, and combat the effects of channel distortions.

In literature, many works have been published on the channel characterization, covering both experimental measurement and computer modeling of indoor and

doi:10.1088/978-0-7503-6049-4ch4

outdoor optical wireless communication (OWC) systems. Most of the optical wireless works in the VLC environment are focusing on indoor channel modeling, while there is little work on outdoor applications, specially multipath mobile V2V channel. The power penalties directly associated with the channel may be separated into two factors, these being optical path loss and multipath dispersion.

In this book, two types of configurations are considered in an optical wireless channel, LoS and NLoS. For directed LoS and tracked configurations, reflections do not need to be taken into consideration, and consequently the path loss is easily calculated from knowledge of the transmitter beam divergence, receiver size and separation distance. However, an NLoS configuration, also known as diffuse systems uses reflections off the surrounding objects' surfaces. Reflections could be seen as unwanted signals or multipath distortions which make the prediction of the path loss more complex. A number of propagation models (ceiling bounce, Hayasaka–Ito and spherical) for LoS and NLoS are introduced in the literature [36], and will be given in brief in this chapter.

4.1.1 Intensity modulation and direct detection (IM/DD)

There are a number of topologies that are generally used for indoor applications. The configurations can be classified according to many factors such as the transmitted angle, the incident angle, and the LoS path between the transmitter and the receiver. The most commonly used technique for the optical communication is the intensity modulation with direct detection (IM/DD). The drive current of an optical source is directly modulated by the modulating signal which in turn varies the intensity of the optical source, as shown in figure 4.1.

The receiver employs a photodiode (PD), with a response which is the integration of tens of thousands of very short wavelengths of the incident optical signal that generates a photocurrent. This photocurrent is directly proportional to the instantaneous optical power incident on it, that is, proportional to the square of received electric field. An IM/DD-based optical wireless system has an equivalent baseband model that hides the high-frequency nature of the optical carrier.

The IM/DD model is shown in figure 4.1. The channel impulse response (CIR) $h(t)$ is a baseband CIR and $n(t)$ is the signal-independent shot noise, modeled throughout the book as the additive white Gaussian noise (AWGN) with a double-sided power spectral density (PSD) of $N_o/2$.

NLoS are subject to the effects of multipath propagation in the same way as radio frequency (RF) systems and these effects are more noticeable. This type of link can suffer from severe multipath-induced performance penalties. Multipath propagation

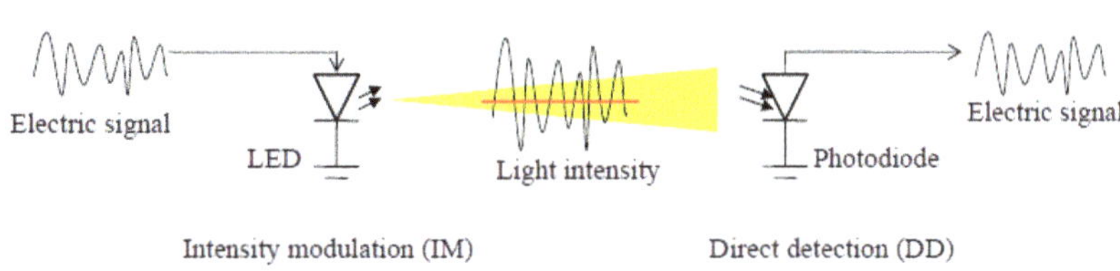

Figure 4.1. An optical intensity, direct detection communications channel.

causes the electric field to suffer from severe amplitude fades on the scale of a wavelength. If the detector size was proportional to one wavelength or less, then the detector will experience multipath fading.

In an IM/DD optical wireless, the received signal can be written as

$$y(t) = \gamma\, x(t) \otimes h(t) + n(t) \tag{4.1}$$

where, $x(t)$ is the transmitted signal, $y(t)$ is the received signal, $h(t)$ is the CIR, γ is the PD responsivity, $\otimes$ is the convolution sign, and $n(t)$ is the noise.

While equation (4.1) is simply a linear filter channel with AWGN, optical wireless systems differ from conventional electrical or radio systems since the instantaneous optical power is proportional to the generated electrical current. $x(t)$ represents the power rather than the amplitude signal. This imposes two constraints on the transmitted signal. Firstly, $x(t)$ must be nonnegative, that is $x(t) \geqslant 0$. Secondly, the eye safety requirements limit the maximum optical transmit power that may be used. Generally, it is the average power requirement which is the most restrictive and hence, the average value of $x(t)$ must not exceed a specified maximum power value $P_{\max}$. This is in difference to the time-averaged value of the signal $|x(t)|^2$, which is the case for the conventional RF channels when $x(t)$ represents amplitude [36]. These differences have a reflective effect on the system design. On conventional RF channels, the signal-to-noise ratio (SNR) is proportional to the average received power, whereas on optical wireless links, it is proportional to the square of the average received optical signal power.

4.2 Indoor VLC channel models

There are a number of topologies that are commonly used for indoor applications. The configurations can be classified according to many parameters such as the degree of directionality of transmitter and receiver and the existence of the LoS path between the transmitter and the receiver. In this chapter, we will give a brief description of the indoor OWC channel models.

4.2.1 LoS propagation model

In this model, the DC gain for a receiver located at a distance of d and angle ϕ with respect to the transmitter can be approximated as [4]

$$h_{\mathrm{LOS}}(0) = \begin{cases} \dfrac{A_r(m_l + 1)}{2\pi d^2}\cos(\phi)T_s(\Psi)g(\Psi)\cos(\Psi) & 0 \leqslant \Psi \leqslant \Psi_c \\ 0 & \text{elsewhere} \end{cases} \tag{4.2}$$

where, m_l is the Lambert's mode number expressing infectivity of the source beam, ϕ is the angle of the radiated power, d is the distance between the transmitter and receiver, $T_s(\Psi)$ is the optical transmission coefficient, $g(\Psi)$ is the optical gain of an ideal non imaging concentrator, A_r is the receiver effective area, Ψ_c is the critical angle and Ψ is the FoV angle, as shown in figure 4.2.

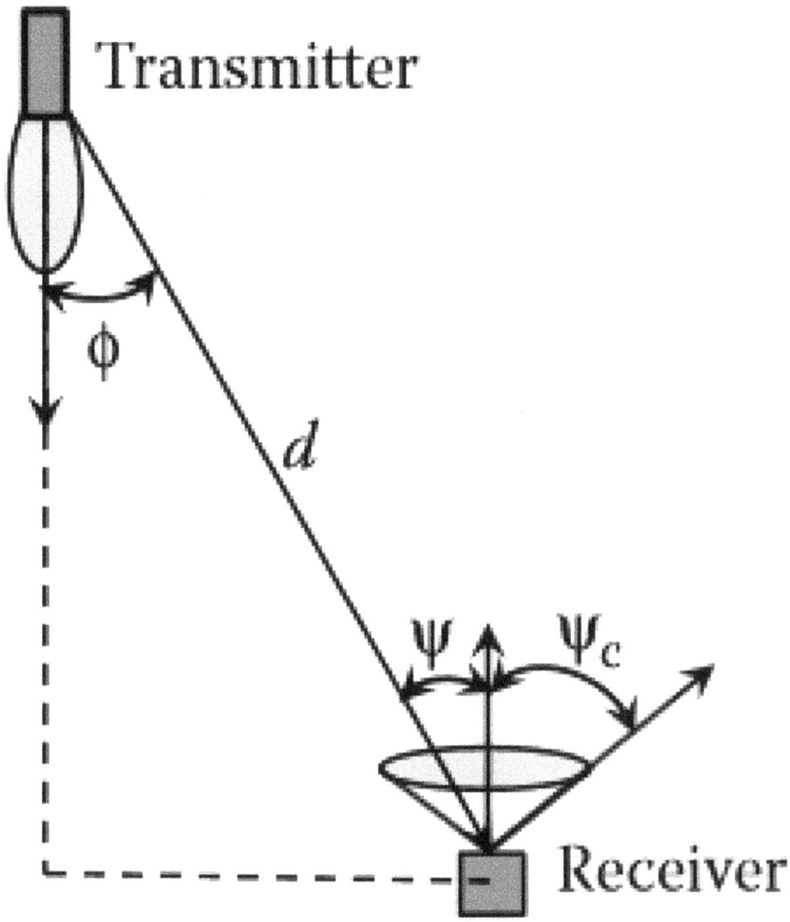

Figure 4.2. Geometry LoS propagation model [36].

In short-distance LOS links, multipath dispersion is seldom a problem and LOS links channel are often modeled as a linear attenuation and delay [5]. The optical LOS links are considered as nonfrequency selective and the path loss depends on the square of distance between the transmitter and the receiver. The impulse response can be expressed as [4]

$$h_{\text{LOS}}(t) = \frac{A_r(m_l + 1)}{2\pi d^2}\cos(\phi)T_s(\Psi)g(\Psi)\cos(\Psi)\delta\left(t - \frac{d}{c}\right) \tag{4.3}$$

where, c is the speed of the light in free space, $\delta(.)$ is the Dirac function.

4.2.2 NLoS propagation model

For NLoS and diffuse links, the optical path loss is more complex to predict since it is dependent on many of factors, such as room dimensions, the reflectivity of the ceiling, walls and objects within the room, and the position and orientation of the transmitter and receiver, window size and place and other physical matters within a room [36]. The received power consists of two components, the LoS power and the NLoS power, and is generally defined as

$$P_r = [H_{\mathrm{LOS}}(0) + H_{\mathrm{NLOS}}(0)]P_t \tag{4.4}$$

where, P_r, *and* P_t represent the received power and transmitted power, respectively.

Mathematically, the impulse response of the optical wireless channel is calculated by integrating the power of all the components arriving at the receiver after multipath propagation [16]. The received signal in the case of the NLoS links consists of various components arriving from different paths, the path length of these components differ in proportion to the room design, hence there is broadening of the pulse. The distribution of the channel gain in decibel (dB) for the LoS component follows a modified gamma distribution, and the channel gain in dB for LoS channels including all reflections follows a modified Rayleigh distribution for most transmitter–receiver distances [6], where, the CIR is given by equation (4.23). The root mean square delay spread is a parameter which is commonly used to quantify the time-dispersive properties of multipath channels, and is defined as the square root of the second central moment of the magnitude squared of the channel impulse response [5], which can be given later in this chapter by equations (4.50) and (4.51), respectively.

4.2.3 Ceiling bounce model

As mentioned before, many models had been proposed and studied for simulating impulse response of indoor Infrared (IR) channel. Carruthers and Kahn presented the ceiling bounce model. This model is commonly used in simulation because of its excellent matching with the measured data and simplicity of the model. This method came up with a closed-form expression for the impulse response assuming the transmitter and the receiver to be collocated in planes parallel to the floor and directed towards the ceiling. In the ceiling bounce model, the multipath IR channel is characterized by only two parameters: the optical path loss and the root-mean-square (RMS) delay spread. The impulse response $h(t)$ of the ceiling bounce model is given by [36]

$$h(t, a) = H(0)\frac{6a^6}{(t + a)^7}u(t) \tag{4.5}$$

where, $u(t)$ is the unit step function, a is the height of the ceiling above the transmitter and receiver.

4.2.3.1 Hayasaka–Ito model

In this model, the diffuse IR channel is analyzed by the ray-tracing technique and the modified Monte Carlo method. Assuming a Lambertian model for the light sources, the impulse response of the channel is expressed as [18]

$$h(t) = h_1(t - T_1) + h_{\mathrm{high}}(t - T_{\mathrm{high}}) \tag{4.6}$$

where, $h_1(t)$ is the impulse response of the primary reflection, $h_{high}(t)$ is that of the higher order reflections, T_1 and T_{high} are the starting times of the rise of the primary and higher order reflections [36].

4.2.4 Spherical model

The spherical model uses with higher order reflections [34] with a gain given by

$$h_{high} = \frac{\eta_{high}}{\tau} e^{\frac{-t}{\tau_e}} \tag{4.7}$$

where, the power efficiency of the diffuse signal η_{high}, the exponential decay time τ_e and the average transmission delay $<t>$ for one reflection are given by [36].

4.3 Outdoor V2V-VLC channel modeling

The development of vehicular communications requires accurate channel models so as to evaluate the system's performance. Channel modeling in a V2V environment is quite complex due to high relative velocities between the transmitter and the receiver in conjunction with a dynamic variation of nodes' position. Another significant characteristic is the high Doppler spread/shift that results in a statistically non stationary channel.

4.3.1 History of channel modeling of vehicular communication

A mathematical model for the V2V-VLC system was presented in [22]. They used a market-weighted headlamp beam pattern considering both the LoS and the NLoS links. They considered the impact of the NLoS path coming from the road surface reflection only. Also, they considered good weather conditions in their analysis. In [12], the authors extended their work in [22], but they assumed a wet surface road and the reflection is not Lambertian. They investigated the system's performance in terms of bit error rate (BER) and the hight of the receiver.

The authors in [20] proposed a measurement-based time variant non-clear sky channel model for the I2V-VLC system. Also, they consider the dynamic characteristics of background radiation to enable more realistic and accurate prediction of the VLC system performance in an outdoor application. Moreover, they introduced a new receiver design with dual-reception and effective ambient-light rejection capabilities which employs the selection diversity technique in order to mitigate the impact of ambient-light noise due to daylight.

In [10], a time function of the V2V channel path loss caused by vehicle mobility had been obtained using the video data collected from cameras mounted on driven vehicles. Computer vision techniques were used to identify the location of the taillights of the transmitting vehicle with respect to the receiving vehicle.

The authors in [30] used a ray-tracing scheme employing commercial light tools software to evaluate channel delay profiles obtained from CIR for multiple LoS links and a less number of NLoS delay taps for crossroad and metropolitan street under the V2V and V2I communication links. They concluded that the delay profile from the V2I link and metropolitan scenario had more dispersive channel characteristics due to the reflection and the diffusion of the visible light.

In [3], a channel model of OWC systems during rainfall was proposed. The authors considered outdoor optical wireless systems where an optical transmitter communicates directly along point-to-point LoS propagation links. They considered the scattering of optical light through raindrops in the propagation channel and derived the probability density function (PDF) of the received signal power.

The authors in [21] evaluated the VLC channel model in the crossroad and metropolitan scenarios based on the practical VLC light sources. They used light tools and CATIA V5 tools to simulate their work.

In [19], the authors used an image sensor instead of PDs as a receiver, a technique considered an attractive solution for outdoor mobile applications. They introduced the pinhole camera model concept to express motion of the VLC transmitters. Moreover, they used the motion models to simulate and analyze VLC channel fluctuation.

4.3.2 2 × 2 MIMO V2V-VLC system model

This section discusses the system model of V2V-VLC in detail. The V2V-VLC system which is shown in figure 4.3 is comprised of two vehicles communicating with each other by utilizing headlights and taillights. Headlights of the first vehicle act as a transmitter and taillights of the second vehicle act as a receiver. Since each vehicle has two headlights and two taillights, a 2 × 2 MIMO link can be considered. The LoS and NLoS paths are considered in this book [14].

4.3.3 The V2V-VLC system considerations

LEDs at the transmitter side and avalanche photodiodes (APDs) at the receiver side are considered. The market-weighted head-lighting pattern source for the

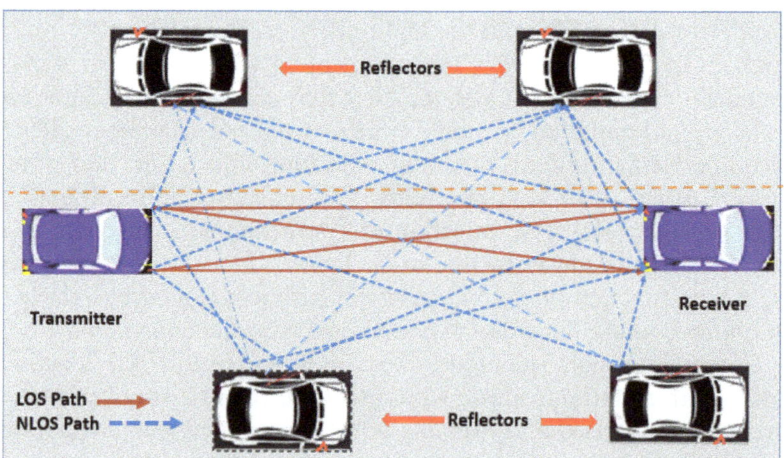

Figure 4.3. (2 × 2) MIMO model for V2V-VLC showing the LoS and NLoS paths. Copyright 2018, IEEE. Reprinted, with permission, from [13].

transmitter and receiver [23] is used in the simulation. These types of lights provide good road illumination, as well as very good anti-glare properties [27]. Moreover, practical outdoor VLC links are considered in this book including parked and passing vehicles acting as reflectors.

Each reflector acts as a receiver when the signal is received from the transmitting vehicle and subsequently as a transmitter when the signal is reflected to the receiving vehicle. Because the reflectance characteristics depend on the reflector's nature and physical state, and it also change with weather conditions, a bidirectional reflectance distribution function (BRDF) is used to model the reflectors in this book. A BRDF model can effectively model the polarimetric signatures of the object surface such as paints or metal [15]. The BRDF is given by [15]

$$f(\theta_{\mathrm{inc}}, \phi_{\mathrm{inc}}, \theta_{\mathrm{ref}}, \phi_{\mathrm{ref}}, \lambda) = \frac{dL_r(\theta_{\mathrm{ref}}, \phi_{\mathrm{ref}})}{dE_i(\theta_{\mathrm{inc}}, \phi_{\mathrm{inc}})} \tag{4.8}$$

where, θ_{inc}, ϕ_{inc}, respectively, denote the incident polar and azimuth angles, θ_{ref}, ϕ_{ref}, respectively, denote the reflected polar and azimuth angles, λ is the wavelength, L_r is the irradiance scattering from the surface, and E_i is the irradiance incident onto the surface.

As outdoor VLC channels undergo fast fading effects, a frequency selective channel model is considered for modeling the system [32].

Referring to the road regulations [26], and by using the data provided in [23], and the headlight radiation pattern shown in figure 4.4, the following points should be considered in V2V-VLC system design:

1. For the effective luminous intensity of 1000 candela (cd), the road width should be about 10 m and the longitudinal distance should be ranging from 80 to 100 m.
2. The safe distance between vehicles traveling at a speed of 80 km/h, and moving in the same lane must be at least 23 m.
3. The safe distance between vehicles in parallel lanes must be at least 3 m.

Due to the above regulations, the V2V-VLC model can have up to six reflecting vehicles within the testing area and only one LOS path.

4.3.4 System description

The block diagram of the V2V-VLC system is shown in figure 4.5. The transmitted serial data stream of the 2×2 MIMO model is converted into two parallel streams using a serial-to-parallel (S/P) converter. Each data stream is intensity modulated and transmitted through LED headlights. At the receiver taillights, PDs convert the light into electrical signals using direct detection.

The received signal can be written as

$$y_r(t) = \gamma\, x(t) \otimes h(t) + n(t) \tag{4.9}$$

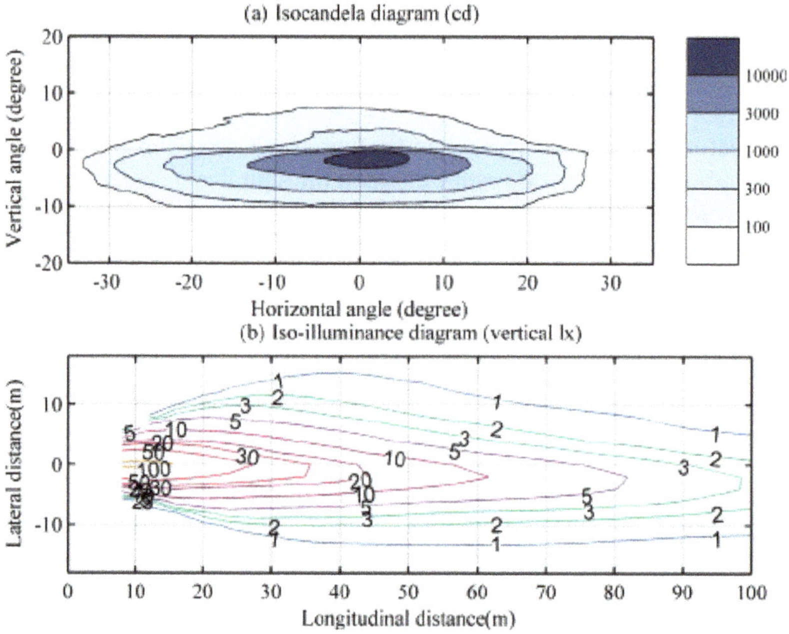

Figure 4.4. Headlight optical radiation pattern. Copyright 2014, IEEE. Reprinted, with permission, from [22].

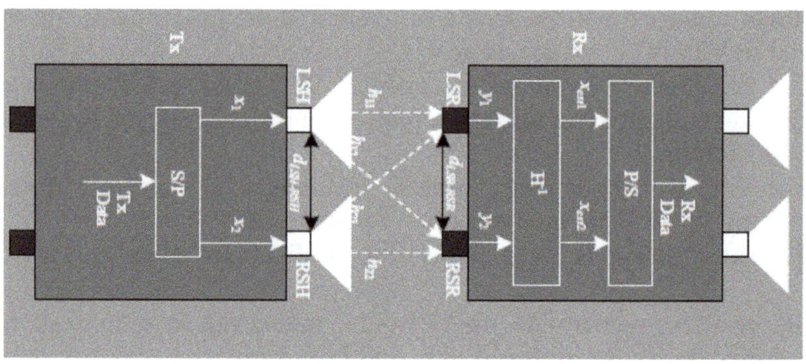

Figure 4.5. Block diagram of a 2 × 2 MIMO link of V2V-VLC system.

where, $y_r(t)$ is the received signal. In vector notation (4.9) is given by

$$\mathbf{y}_r = \gamma\,\mathbf{H}\,\mathbf{x} + \mathbf{n} \tag{4.10}$$

where, $\mathbf{x}$ and $\mathbf{y}_r$ are transmitted and received vectors, respectively, $\mathbf{H}$ is the channel matrix which consists of the LoS and the NLoS components, and $\mathbf{n}$ is the noise vector.

The channel matrix is given by

$$\mathbf{H} = \begin{bmatrix} h_{11} & h_{12} \\ h_{21} & h_{22} \end{bmatrix} \tag{4.11}$$

Referring to figure 4.5 x_1 *and* x_2 are the transmitted signals from first and second transmitters s_1 and s_2, respectively, y_1 and y_2 are the received signals by both receivers $r_1$and r_2, while n_1, and n_2 represent the Gaussian noise at receiver r_1, and r_2, respectively.

4.4 Channel impulse response and transfer function

The total CIR of the V2V-VLC model, $h(t)$ is given by

$$h(t) = h(t)_{\mathrm{LOS}} + h(t)_{\mathrm{NLOS}} \tag{4.12}$$

By performing fast Fourier trransform (FFT) on (4.12) we get

$$H(\omega) = H(\omega)_{\mathrm{LOS}} + H(\omega)_{\mathrm{NLOS}} \tag{4.13}$$

4.4.1 Impulse response and transfer function of the LoS component

The 2×2 impulse response matrix H_{LOS} between the transmitters and the receivers for the LoS path can be written as

$$H_{\mathrm{LOS}} = \begin{bmatrix} h_{11} & h_{12} \\ h_{21} & h_{22} \end{bmatrix} \tag{4.14}$$

Any element in H_{LOS} is expressed as

$$h_{rs}(t) = G_{rs}\, \delta\, (t - \tau_{rs}) \qquad r = 1, 2, \quad and \quad s = 1, 2. \tag{4.15}$$

where, G_{rs} denotes the gain factor accounting for the losses and gains between the source s and the receiver r, $\tau_{rs} = R_{rs}/c$, denotes the time delay, R_{rs} denotes the distance between the source s and receiver r, c is the speed of light, and $\delta()$ denotes the impulse function.

The gain factor, G_{rs}, is given by [29]

$$G_{rs} = \frac{\cos(\phi_{rs})\cos(\theta_{rs})\, A_r\, T_s(\theta_{rs})\, g(\theta_{rs})}{\pi\, R_{rs}^2}. \tag{4.16}$$

where, ϕ_{rs} denotes the emitting angle between source s and receiver r, θ_{rs} denotes the incident angle between source s and receiver r, A_r denotes the effective area of the receiver, $T_s(\theta_{rs})$ denotes the signal transmission coefficient of an optical filter, and $g(\theta_{rs})$ denotes the concentrator gain. Figure 4.6 shows the relevant angles for the transmitter, the receiver, and the reflector.

In the frequency domain, the LoS transfer function of (4.15) is given by

$$H_{rs}(\omega) = G_{rs}\, e^{-j\,\omega\,\tau_{rs}} \tag{4.17}$$

4.4.2 Impulse response and transfer function of the NLoS component

Let each reflector surface can be divided into N small surfaces (subsurfaces) elements numbered from (1, 2, … , N) with equal areas A_i and equal reflectivity factors $\rho_i < 1$ ($i = 1, 2, \ldots, N$), as shown in figure 4.7.

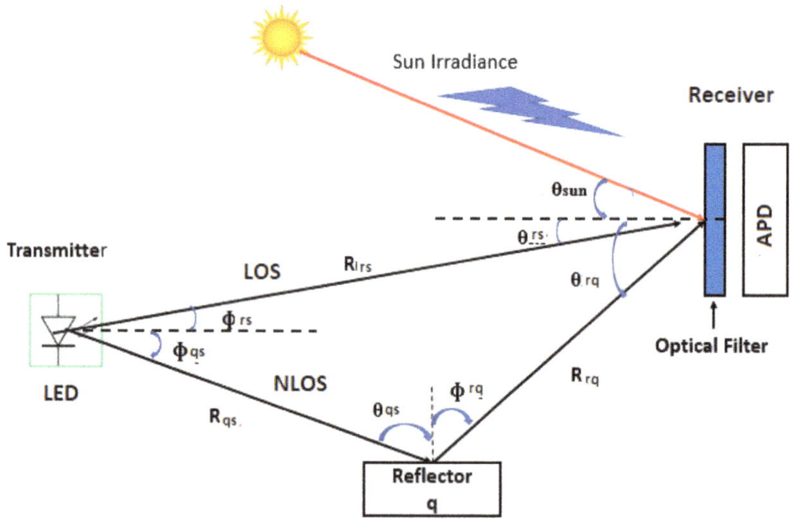

Figure 4.6. Emitting and incident angles and other related parameters.

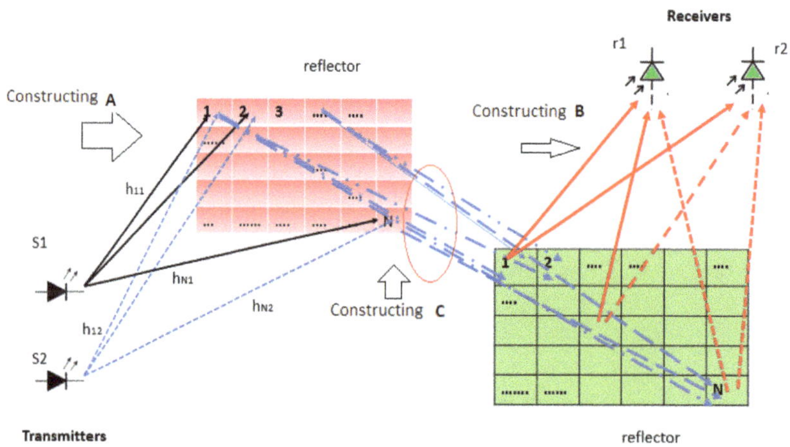

Figure 4.7. Illustration of how the NLoS path's matrices (**A**, **B**, and **C**) are constructed.

When the transmitted light incidents on the reflector surface, the surface element will be considered as a receiver. Thus, the impulse response matrix A between the transmitters ($s = 1, 2$) and the subsurfaces of any reflector, as shown in figure 4.7, is given by

$$A = \begin{bmatrix} h_{11} & h_{12} \\ h_{21} & h_{22} \\ \cdots & \cdots \\ h_{N1} & h_{N2} \end{bmatrix} \tag{4.18}$$

Each element h_{is} in the matrix A is given by:

$$h_{is}(t) = G_{is}\,\delta\,(t - \tau_{is}) \qquad s = 1, 2, \text{ and } i = 1, 2, \dots, N \tag{4.19}$$

The gain factor between the transmitter and any subsurface G_{is} is given by [29]:

$$G_{is} = \frac{\cos(\phi_{is})\cos(\theta_{is})\,A_i}{\pi\,R_{is}^2} \tag{4.20}$$

where, ϕ_{is} denotes the emitting angle between source s and reflector subsurface i, θ_{is} denotes the incident angle between source s and reflector subsurface i, R_{is} denotes the distance between source s and reflector subsurface i, and A_i denotes the effective area of the reflector subsurface i.

The transfer function of (4.19) $H_{is}(\omega)$ is given by (4.17), with r replaced by i.

Assuming a high order reflection system, i.e. the transmitter s emits the light to the reflector 1, which then reflects it to reflector 2 and from 2 to 3 and so on. This process repeats until the last reflector radiates the power to the receiver r directly. In every reflection, the optical power is reduced by the reflectivity factor ρ_i.

The impulse response matrix B between the receiver r $(r = 1, 2)$, and the last reflector (which has N subsurfaces) is given by

$$B = \begin{bmatrix} h_{11} & h_{12} & \cdots & h_{1N} \\ h_{21} & h_{22} & \cdots & h_{2N} \end{bmatrix} \tag{4.21}$$

The gain factor G_{ri}, and the transfer function $H_{ri}(\omega)$ of each element in (4.21) are given by (4.15) and (4.17), respectively, but with the proper replacements of i instead of s.

The corresponding impulse response matrix between any two reflectors (each with N subsurfaces) $\mathbf{C}$ is

$$\mathbf{C} = \begin{bmatrix} h_{11} & h_{12} & \cdots\cdots & h_{1N} \\ h_{21} & h_{22} & \cdots\cdots & h_{2N} \\ \cdots & \cdots & \cdots\cdots & \cdots \\ h_{N1} & h_{N2} & \cdots\cdots & h_{NN} \end{bmatrix} \tag{4.22}$$

where, h_{ki}, and the gain G_{ki}, are given by (4.15) and (4.16), respectively $(i, k = 1, 2, \dots, N$, $(i$ *and* $k)$ are the subsurface's index of the first and second reflector, respectively. Also, the transfer function $H_{ki}(\omega)$ is given by (4.17), after proper replacement of k with r, and i with s. Figure 4.7 illustrates how the NLoS path's matrices (**A, B**, and **C**) are constructed.

The total impulse response $h_{\mathrm{NLOS}}(t)$ for the NLoS components is the sum of the contributions from the reflections up to a given order L [17]

$$h_{\mathrm{NLOS}}(t) \approx \sum_{l=1}^{L} h_{\mathrm{NLOS}}^{(l)}(t) \tag{4.23}$$

As an example, the first-order reflection system (i.e. one reflector with N subsurfaces $(i = 1, 2, \dots, N)$ has an impulse response given by [29]

$$h_{\mathrm{NLOS}}^{(1)}(t) = \sum_{i=1}^{N} \rho_i h_{ri}(t) \otimes h_{is}(t) \tag{4.24}$$

The second-order reflection system (i.e. two reflectors with N subsurfaces each $(i, k = 1, 2, \ldots, N)$ is given by

$$h_{\mathrm{NLOS}}^{(2)}(t) = \sum_{i=1}^{N}\sum_{k=1}^{N} \rho_k \rho_i h_{kr}(t) \otimes h_{ki}(t) \otimes h_{is}(t) \tag{4.25}$$

The third-order reflection system (i.e. three reflectors with N subsurfaces each $(i, k, l = 1, 2, \ldots, N)$, ($l$ is the subsurface's index of the third reflector) is given by

$$h_{\mathrm{NLOS}}^{(3)}(t) = \sum_{l=1}^{N}\sum_{i=1}^{N}\sum_{k=1}^{N} \rho_k \rho_i\ \rho_l h_{kr}(t) \otimes h_{li}(t) \otimes h_{kl}(t) \otimes h_{is}(t) \tag{4.26}$$

and so on.

Many previous works such as [7] and [35] used recursive and iterative formulas to approximate (4.23). By the end, they got an exponential decay model for the VLC outdoor environment which is given by,

$$h_{\mathrm{NLOS}}(t, \tau) = \frac{1}{\tau}\, e^{\frac{-t}{\tau}} \tag{4.27}$$

where, the time constant $\tau = 2\ \tau_{\mathrm{RMS}}$ and τ_{RMS} is the channel RMS delay spread.

4.4.3 Derivation of NLoS transfer function

The frequency domain version of (4.18) is given by

$$A(\omega) = [h_{i1}(\omega) \quad h_{i2}(\omega)] \tag{4.28}$$

where, $h_{i1}(\omega)$ and $h_{i2}(\omega)$ are two vectors that indicate the transfer functions of the first and second transmitter and the reflector subsurfaces, respectively.

Any vectors in (4.28) can be expanded to

$$h_{is}(\omega) = [G_{1s}\, e^{-j\,\omega\,\tau_{1s}}\ G_{2s}\, e^{-j\,\omega\,\tau_{2s}}\ \cdots\ G_{Ns}\, e^{-j\,\omega\,\tau_{Ns}})]^T \qquad s = 1, 2 \tag{4.29}$$

Similarly, the frequency domain version of (4.21) is given by

$$B(\omega)^T = [h_{1i}(\omega) \quad h_{2i}(\omega))] \tag{4.30}$$

where, $h_{ri}(\omega)$, and $h_{2i}(\omega)$ are two vectors that indicate the transfer functions of the first and second receiver with the last reflector surface.

The vectors in (4.30) can be expanded to

$$h_{ri}(\omega)^T = [G_{r1}\, e^{-j\,\omega\,\tau_{r1}}\ G_{r2}\, e^{-j\,\omega\,\tau_{r2}}\ \cdots\ G_{rN}\, e^{-j\,\omega\,\tau_{rN}})]^T \qquad r = 1, 2 \tag{4.31}$$

The $N \times N$ frequency domain version of (4.22) can be written as

$$C(\omega) = \begin{bmatrix} G_{11}\, e^{-j\,\omega\,\tau_{11}} & G_{12}\, e^{-j\,\omega\,\tau_{12}} & \cdots & G_{1N}\, e^{-j\,\omega\,\tau_{1N}} \\ G_{21}\, e^{-j\,\omega\,\tau_{21}} & G_{22}\, e^{-j\,\omega\,\tau_{22}} & \cdots & G_{2N}\, e^{-j\,\omega\,\tau_{2N}} \\ \cdots & \cdots & \cdots & \cdots \\ G_{N1}\, e^{-j\,\omega\,\tau_{N2}} & G_{N2}\, e^{-j\,\omega\,\tau_{N2}} & \cdots & G_{NN}\, e^{-j\,\omega\,\tau_{NN}} \end{bmatrix} \tag{4.32}$$

To include the reflectivity factors ρ_i of the surfaces, we define the $N \times N$ reflectivity matrix as follows

$$G_\rho = diag[\rho_1 \;\; \rho_2 \;\; \cdots \;\; \rho_N] \tag{4.33}$$

If all the reflection coefficients are assumed to be equal i.e, $(\rho_1 = \rho_2 = \cdots = \rho_N)$.

To have a closed-form expression for the NLoS link transfer function, we have to find the FFT of (4.24), (4.25) and (4.26) till the lth reflection order, then applying Carl Neumann's result on matrices as follows, for first-order reflection

$$H^{(1)}{}_{\text{NLOS}}(\omega) = B(\omega)\; G_\rho\; A(\omega) \tag{4.34}$$

For second-order reflection

$$H^{(2)}{}_{\text{NLOS}}(\omega) = B(\omega)\; G_\rho\; C(\omega)\; G_\rho\; A(\omega) \tag{4.35}$$

For third-order reflection

$$H^{(3)}{}_{\text{NLOS}}(\omega) = B(\omega)\; G_\rho\; (C(\omega)\; G_\rho)^2\; A(\omega) \tag{4.36}$$

For the lth order reflection

$$H^{(l)}{}_{\text{NLOS}}(\omega) = B(\omega)\; G_\rho\; (C(\omega)\; G_\rho)^{l-1}\; A(\omega) \tag{4.37}$$

The total NLoS transfer function is given by the infinite series

$$\begin{aligned} H_{\text{NLOS}}(\omega) &= \sum_{l=1}^{\infty} H^{(l)}_{\text{NLOS}}(\omega) \\ &= H^{(1)}{}_{\text{NLOS}}(\omega) + H^{(2)}{}_{\text{NLOS}}(\omega) + H^{(3)}{}_{\text{NLOS}}(\omega) + \cdots + H^{(l)}{}_{\text{NLOS}}(\omega) + \cdots \\ &= B(\omega)\; G_\rho\; A(\omega) + B(\omega)\; G_\rho\; C(\omega)\; G_\rho\; A(\omega) + B(\omega)\; G_\rho\; (C(\omega)\; G_\rho)^2 \\ &\quad A(\omega) + \cdots + B(\omega)\; G_\rho\; (C(\omega)\; G_\rho)^{l-1}\; A(\omega) + \cdots \\ &= B(\omega)\; G_\rho \sum_{m=0}^{\infty} (C(\omega)\; G_\rho)^m\; A(\omega) \end{aligned} \tag{4.38}$$

Applying Carl Neumann's result on matrices ($(I_n - F)^{-1} = \sum_{m=0}^{\infty} F^m$), then, the expression in (4.38) can be summed up to

$$\sum_{m=0}^{\infty} (C(\omega)\; G_\rho)^m = (I - C(\omega)\; G_\rho)^{-1} \tag{4.39}$$

where, **I** denotes the unity matrix.

Rewrite (4.38) to get the final expression for the NLoS transfer function which is given by the matrix product as

$$H_{\mathrm{NLOS}}(\omega) = B(\omega)\, G_{\rho}\, (I - C(\omega)\, G_{\rho})^{-1}\, A(\omega) \tag{4.40}$$

Equation (4.40) gives a closed-form expression for the NLoS path, where the infinite number of reflections shown by (4.38) becomes not an issue.

The impulse response $h_{\mathrm{NLOS}}(t)$ can then be obtained by performing inverse FFT (IFFT) on the transfer function $\mathbf{H}_{\mathrm{NLOS}}(\omega)$. Therefore, the total impulse response of the suggested model $h(t)$ can then be obtained by performing IFFT on the transfer function given by (4.13).

4.5 Performance analysis of the V2V-VLC channel model

4.5.1 The received optical power and SNR

The total received optical power P_r can be calculated as follows:

$$P_r = P_r^{\mathrm{LOS}} + P_r^{\mathrm{NLOS}} \tag{4.41}$$

The received optical power from the LOS path is given by

$$P_r^{\mathrm{LOS}} = \begin{cases} \left(\sum_{r=1}^{2} \sum_{s=1}^{2} \dfrac{I_s(\alpha_s, \beta_s)\, A_r}{(\mathrm{LER})\, R_{rs}^2} \cos(\phi_{rs}) \cos(\theta_{rs}) \right) & \text{if } 0 \leqslant \theta_{rs} \leqslant \Psi_{rs} \\ 0, & \text{if } \theta_{rs} > \Psi_{rs} \end{cases} \tag{4.42}$$

where, $I_s(\alpha_s, \beta_s)$ is the luminous intensity of the transmitter s from the direction (α_s, β_s) and its measured in candela (cd), (α_s, β_s) are the vertical and horizontal angles of the headlight, respectively, Ψ is the receiver's FoV, LER is luminous efficacy of radiation of a high power LED.

The received optical power from the NLoS paths is given by

$$P_r^{\mathrm{NLOS}} = \begin{cases} \left(\sum_{r=1}^{2} \sum_{s=1}^{2} \sum_{q=1}^{Q} \dfrac{I(\alpha_s, \beta_s) \cos(\theta_{qr})\, A_q\, A_r}{(\mathrm{LER})\, R_{qs}^2 \left(R_{rq}^2 + h_i^2\right)} \rho_q \cos(\phi_{rq}) \cos(\theta_{rq}) \right) & \text{if } 0 \leqslant \theta_{rq} \leqslant \Psi_{rq} \\ 0, & \text{if } \theta_{rq} > \Psi_{rq} \end{cases} \tag{4.43}$$

where, Q is the total number of the reflectors, θ_{qs} denotes the incident angle between reflector q and receiver r, A_q denotes the reflector effective area, h_i denotes the hight of the receiver, R_{qs} denotes the distance between the transmitter and the reflector, R_{rq} denotes the distance between the reflector and the receiver, ϕ_{rq} denotes the emission angle between the reflector and receiver, θ_{rq} denotes the incident angle between the receiver and the reflector, and ρ_q denotes the reflection factor of reflector q. The illustration of these parameters can be seen in figure 4.6.

In this chapter, we assume the on–off keying (OOK) modulation scheme and both transmitters have the same power. Thus, at the receiver, the SNR is given by [7, 11]

$$SNR = \frac{(\gamma\, P_r)^2}{\sigma_n^2} \tag{4.44}$$

where, σ_n^2 is the noise power given by (2.7).

4.5.2 Calculating the BER

BER is the probability that an error may occur in a bit in the pulse train, i.e., a '1' bit turns into a '0' bit or vice versa.

$$BER = P(0)\,P(e/0) + P(1)\,P(e/1) \tag{4.45}$$

where, $P(0)$ and $P(1)$ are the probability of transmit '0' and '1', respectively, $P(e/0)$ and $P(e/1)$ are the conditional probability for receiving '1' while '0' was transmitted and receiving '0' while '1' was transmitted, respectively.

If OOK-non-return-to-zero (NRZ) transmission is considered and random noise Gaussian distribution. Moreover, let both symbols have the same variances and they have the same probability to be transmitted (equally likely) i.e. $P(0) = P(1) = 1/2$, in this case, the threshold level is set to be halfway between the symbols, i.e at $V_m/2$, as shown in figure 4.8.

The conditional probability functions can be calculated as

$$P(e/1) = \frac{1}{\sigma\sqrt{2\pi}} \int_{-\infty}^{\frac{V_m}{2}} e^{\frac{-(x-\mu_1)^2}{2\sigma^2}} dx \tag{4.46}$$

$$P(e/0) = \frac{1}{\sigma\sqrt{2\pi}} \int_{\frac{V_m}{2}}^{\infty} e^{\frac{-(x-\mu_0)^2}{2\sigma^2}} dx \tag{4.47}$$

where, $\mu_0 = 0$ and $\mu_1 = V_m$ for the NRZ signal.

Substituting (4.46) and (4.47) in (4.45)

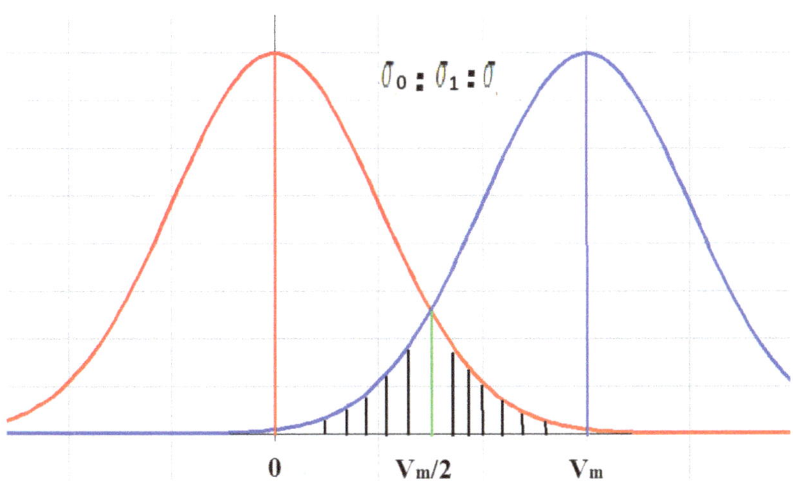

Figure 4.8. Conditional PDF with OOK modulation.

$$
\begin{aligned}
BER_{OOK} &= \frac{1}{2\sigma\sqrt{2\pi}}\left(\int_{-\infty}^{\frac{V_m}{2}} e^{\frac{-(x-V_m)^2}{2\sigma^2}}dx + \int_{\frac{V_m}{2}}^{\infty} e^{\frac{-(x-0)^2}{2\sigma^2}}dx\right) \\
&= \frac{1}{2}\left(\frac{1}{2}erfc\left(\frac{1}{2\sqrt{2}}\sqrt{SNR}\right) + \frac{1}{2}erfc\left(\frac{1}{2\sqrt{2}}\sqrt{SNR}\right)\right) \\
&= \frac{1}{2}erfc\left(\frac{1}{2\sqrt{2}}\sqrt{SNR}\right) \\
&= \frac{1}{2}erfc\left(\frac{1}{2\sqrt{2}}\sqrt{\frac{(\gamma P_r)^2}{\sigma_n^2}}\right)
\end{aligned}
\tag{4.48}
$$

where, *erfc*(.) is the complementary error function used to calculate the tail probability of the standard Gaussian distribution, and is given by

$$
erfc(x) = \frac{2}{\sqrt[2]{\pi}}\int_x^{\infty} e^{-y^2}\,\mathrm{d}y \tag{4.49}
$$

4.5.3 Time dispersion parameters for the channel

By using the derived CIR given by (4.12), we can calculate several channel parameters such as RMS delay spread, mean excess delay, frequency correlation function, coherence time, and coherence bandwidth.

The RMS delay spread and mean excess delay are given by [28]

$$
\tau_{\mathrm{RMS}} = \sqrt{\frac{\int_0^{\infty} (t-\bar{\tau})^2\, h^2(t)dt}{\int_0^{\infty} h^2(t)dt}} \tag{4.50}
$$

where, $\bar{\tau}$ is the mean excess delay and it is given by

$$
\bar{\tau} = \frac{\int_0^{\infty} t\, h^2(t)dt}{\int_0^{\infty} h^2(t)dt} \tag{4.51}
$$

For the V2V-VLC environment, the frequency correlation function $H(\triangle\omega)$ can be obtained from the CIR as [25]

$$
H(\triangle\omega) = \int_{-\infty}^{\infty} h(t)\, e^{-j\,\Delta\omega\, t}\, dt \tag{4.52}
$$

Once the frequency correlation function is calculated, the coherence bandwidth B_c for a VLC channel can be found as [33]

$$
B_c = |min \triangle (\omega)| \tag{4.53}
$$

Since V2V-VLC system is good for short-distance communications. The effective velocity of both the transmitter and the receiver ($V_{\mathrm{eff}} = \sqrt{V_s^2 + V_r^2}$) compared to the velocity of light is very small, thus, a correlation level of 0.9 is typically chosen [33], i.e $|H \triangle (\omega)| = 0.9$.

4.6 Simulation and results

MATLAB software is used to simulate the V2V-VLC system. The simulation parameters are shown in tables 2.4 and 4.1. Multipath propagation is considered. We assume all reflector surfaces to have the same reflectively factor and same areas. We consider the moving speed of 30 km h^{-1} for the transmitter and 25 km h^{-1} for the receiver (whether they move in the same direction, or in the opposite direction). We use Philips Ultinon LED 12985BWX2 as transmitter [1]. Also, we use an APD model (Si) Hamamatsu S8664-1010 as a receiver [2]. We analyze the system performance considering various parameters.

Figure 4.9 shows the CIR of the LoS and the NLoS components when the transmitter and the receiver are moving in the same directions, but with different velocities. The CIR of the LoS path is represented by an impulse, while the CIR of the NLoS components has an exponential decay behavior. Moreover, figure 4.9 shows that, as the distance between the transmitter and the receiver increases, the delay increases but the channel gain decreases. The low gain shown in figure 4.9 for both paths is due to the nature of V2V-VLC system where the channel is changing very fast. In addition, both the transmitter and the receiver are moving affecting both the angle of emitting, as well as the angles of incidence. The importance of the

Table 4.1. Simulation parameters.

Parameter	Value
Number of reflectors	2–6
Modulation type	OOK
Distance between the reflectors	3–80 m
Distance between the transmitter and the receiver	20–120 m
Receiver area	4 cm^2
Total area of the reflector	0.5 m^2
LED luminance intensity	1000–2500 cd
Transmitted and received angles	5°–45°
Responsivity of PD γ	0.54 A/W
Reflectivity ρ	0.8
Gain of the optical filter T(Ψ)	1
Gain of the optical concentrator g	5
Frequency correlation level	0.9
Speeds of the transmitter and receiver, respectively	30 km h^{-1}, 25 km h^{-1}
Data rate	35 Mbps
Receiver height	70 cm

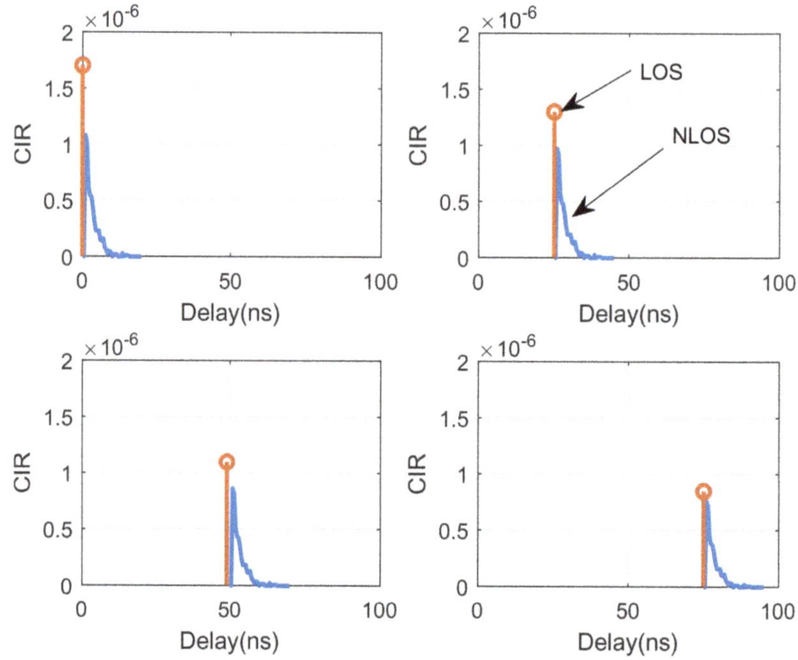

Figure 4.9. Channel impulse response of the LoS and NLoS components when vehicles are moving in the same direction.

channel gain in the V2V-VLC system rising from it determines the achievable SNR for fixed transmitting power. Figure 4.10 shows the transfer function of the V2V-VLC channel. We use (4.40) to simulate the NLoS paths for up to six reflectors, while we use (4.16) and (4.17) to simulate the gain of the LoS path. From figure 4.10, we see that the gain of the channel depends not only on frequency but also on other factors such as emitting angle, incident angle, and the reflectivity of the surrounding objects. Also, figure 4.10 shows the difference between LoS gain and NLoS gain is inversely proportional to a number of reflections. This means that we need a high number of reflections to make the difference close to zero. The exponential decay of the CIR for the diffusion path shown in figure 4.9 is obtained after performing IFFT for the transfer function shown in figure 4.10. Smoothing operation has to be done for the transfer function (such as raised cosine window) before applying the IFFT. The simulation results show acceptable similarity with the mathematical model for CIR given by (4.27) as shown in figure 4.11.

Figure 4.12 shows the received power versus the distance between the vehicles for up to six reflectors. The received power is inversely proportional to the distance. Furthermore, it can be seen that the contribution by the high order reflections ($l > 2$) are not dominant. The dominant components are coming from the LoS path as well as from the first and second reflections. This is because of the scattering property of the light and the absorption by the reflectors and other surrounding objects reduce the received power strength. BER versus different values for R_{rs} at a data rate of 35 Mbps is shown in figure 4.13. It can be seen that, as the distance increases, the system

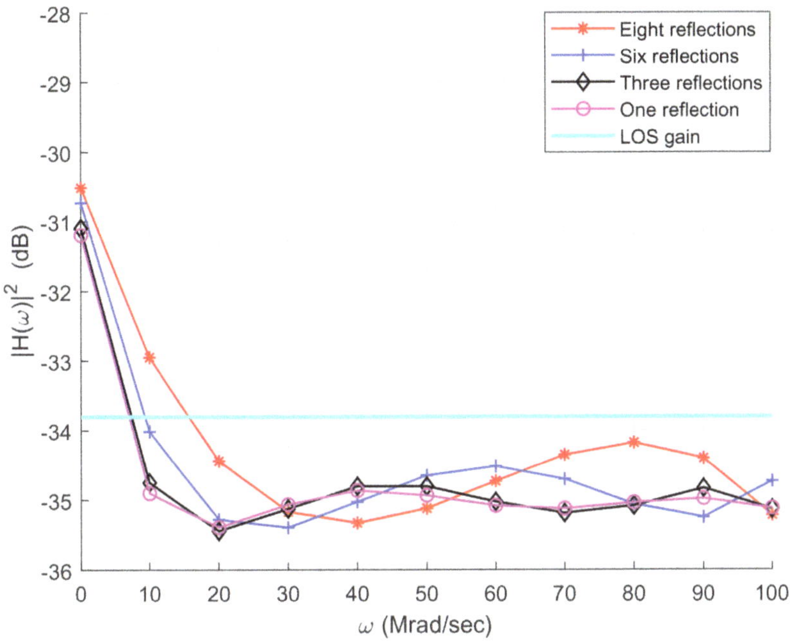

Figure 4.10. The LoS and NLoS paths transfer functions.

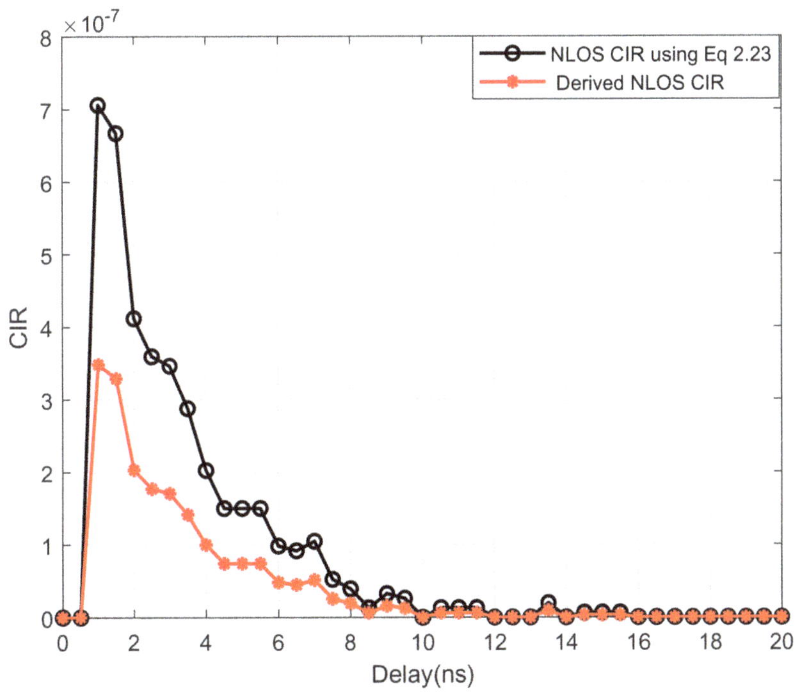

Figure 4.11. Comparison between derived NLoS CIR and NLoS CIR given by equation (2.23).

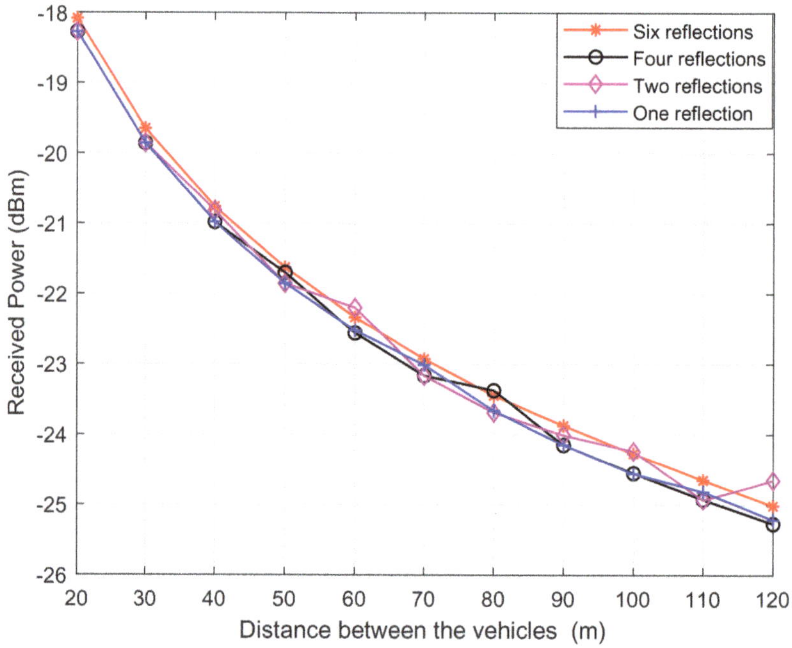

Figure 4.12. Received power versus the distance between vehicles.

BER performance decreases. Also, we can see from figure 4.13 that, the value of the BER degrades from 10^{-6} at $R_{rs} = 30$ m to 3.5×10^{-5} at $R_{rs} = 80$ m. This happens because when R_{rs} changes from 30 to 80 m, the received optical power reduces by more than 3 times (e.g., it decreases from –24 dBm to –20 dBm) according to figure 4.12.

The investigated model shows better performance than the results obtained in [22] and[12], since we consider passing and parking cars as reflectors. Also, figure 4.13 emphasizes that the high order number of reflections will not cause significant difference in the performance of the system.

Figure 4.14 shows the PDF of the delay spread for the V2V-VLC channel. It can be seen that the best fitted distribution for the RMS of the channel in the case of the same direction movement is similar to Rician distribution. The mean value of the delay spread of V2V-VLC model is about 12 ns, while the authors in [31] found it to be 10–15 ns as shown in table 4.2. This is attributed to the fact that the distance is short while the signal speed is very high in V2V-VLC systems, which produce a dispersive channel. In addition, motion introduces additional Doppler spreading and shifting.

Figures 4.15, and 4.16 depict the average of 50% and 90% coherence time of the suggested channel when the transmitter and the receiver are moving in the same direction, and in opposite direction, respectively. There is a decreasing trend in coherence time with effective speed for both directions, as expected, which is consistent with the Doppler spread analysis. The VLC channel shows a larger coherence time than the values reported for a V2V-RF channel in the urban

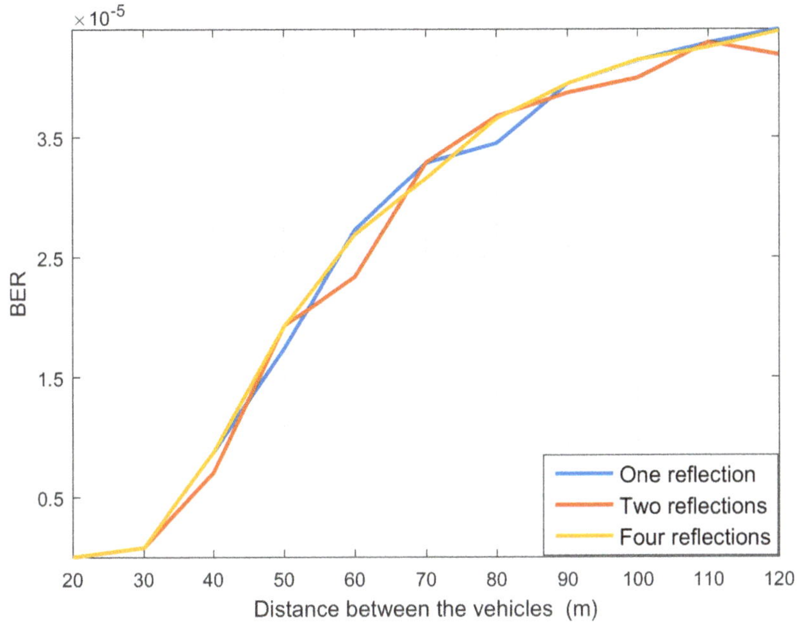

Figure 4.13. BER versus the distance between vehicles moving in the same direction.

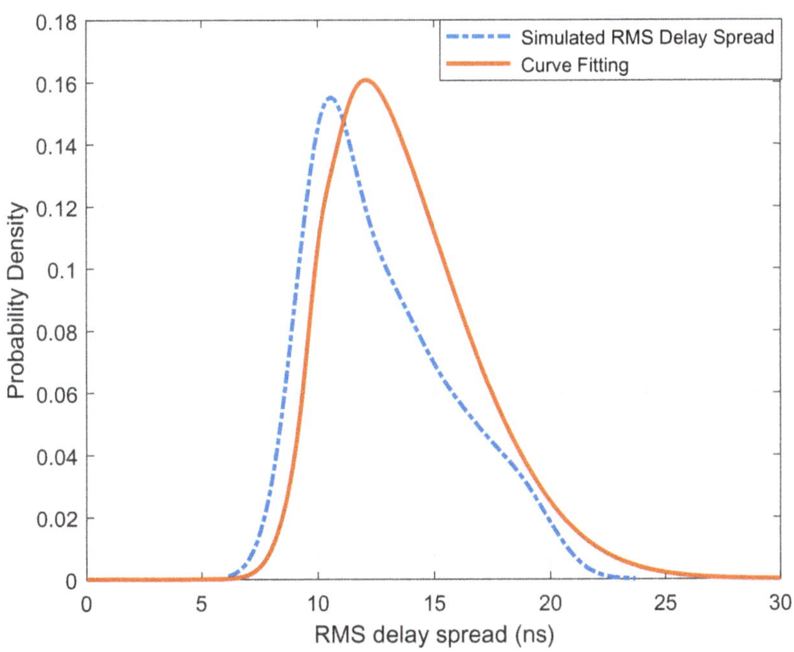

Figure 4.14. RMS delay spread for one reflection system.

Table 4.2. RMS delay spread comparison.

Reference	Delay spread (ns)
V2V [31]	10
V2I [31]	15
Infrared channels [5]	2
For this work when vehicles move in the same direction	12

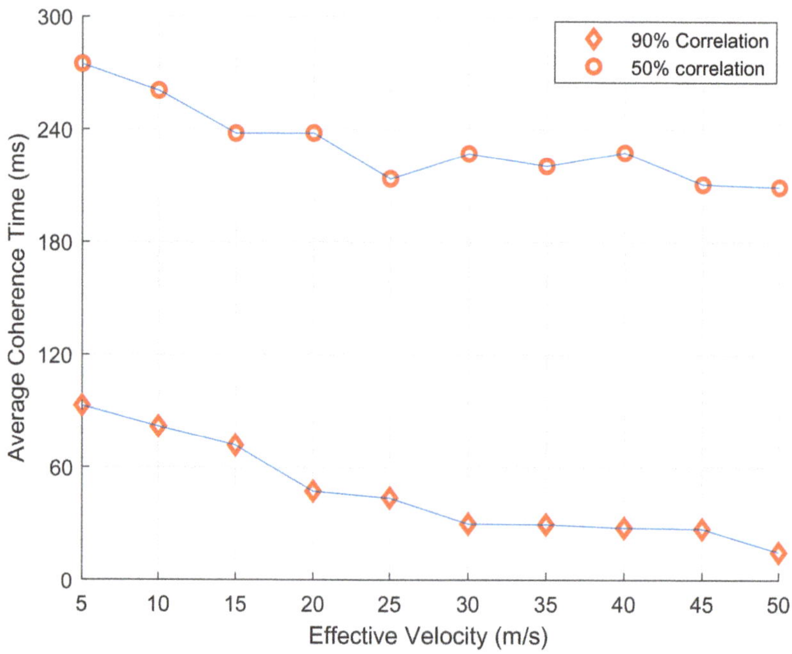

Figure 4.15. Channel coherence time when both vehicles move in the same direction.

scenarios [9, 24]. Table 4.3 compares the median coherence time of (VLC and RF) obtained from this work and several previous works. Moreover, the V2V-VLC model shows that 2 × 2 MIMO-VLC links have much slower channel time variation compared to RF V2V links, and other VLC models. Also, the V2V-VLC channel coherence time is found to be at least an order of magnitude larger than that of RF channels, indicating that the V2V-VLC channels are much stable.

Here, the coherence time was computed directly from received signals instead of assuming a reciprocal time-frequency relationship. The fact that the coherence time of the VLC system is high, specially when both vehicles move in the same direction, is attributed to the very high frequency of light. The movement in the opposite direction appears to have lower channel coherence time than the movement in the same direction. This is because the effective speed of both vehicles will be higher, and the dominant path is the NLoS only. We can suggest that the V2V-VLC channel can

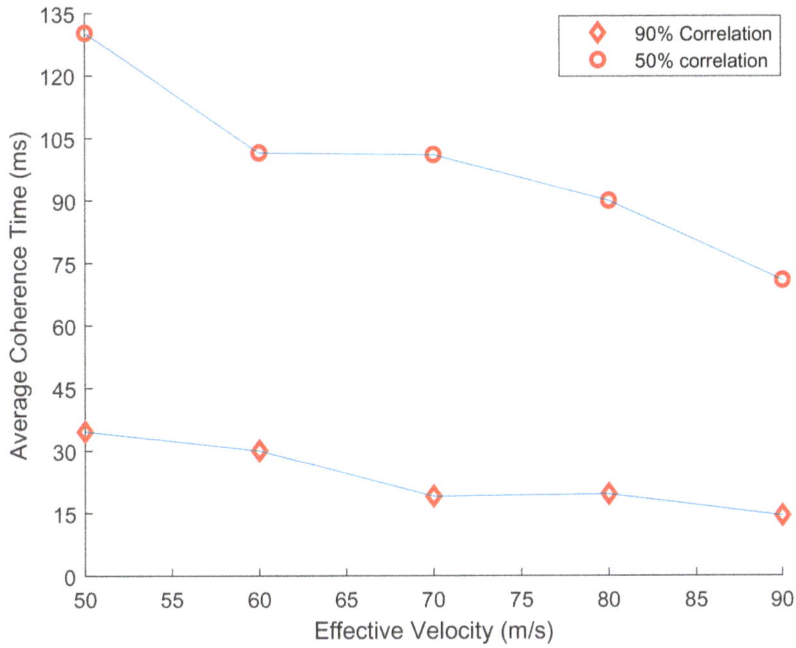

Figure 4.16. Channel coherence time when both vehicles move in the opposite direction.

Table 4.3. Median coherence time comparison (units in (ms)).

Channel	50%	90%
RF channel [9]	103	3.6
VLC channel [8]	164.7	46.8
VLC channel [10]	367	33
For this work when vehicles move in the same direction	220	50
For this work when vehicles move in the opposite direction	100	25

be considered as a selective fading channel. Note that the simulated coherence time does not continue to decrease monotonically with distance since VLC signal strength varies rapidly, it is plausible that small perturbations in the environment could lead to relatively large changes in the channel gain.

The coherence bandwidth values as a function of separation distances are shown in figure 4.17, where the instantaneous coherence bandwidth is inversely proportional to the separation distance. The highest coherence bandwidth values are available for the lowest RMS delay spread values, and higher values can be seen for the same direction movement. The coherence bandwidth values range between 3 and 7 MHz for the opposite direction movement, and 4–25 MHz for the same direction movement.

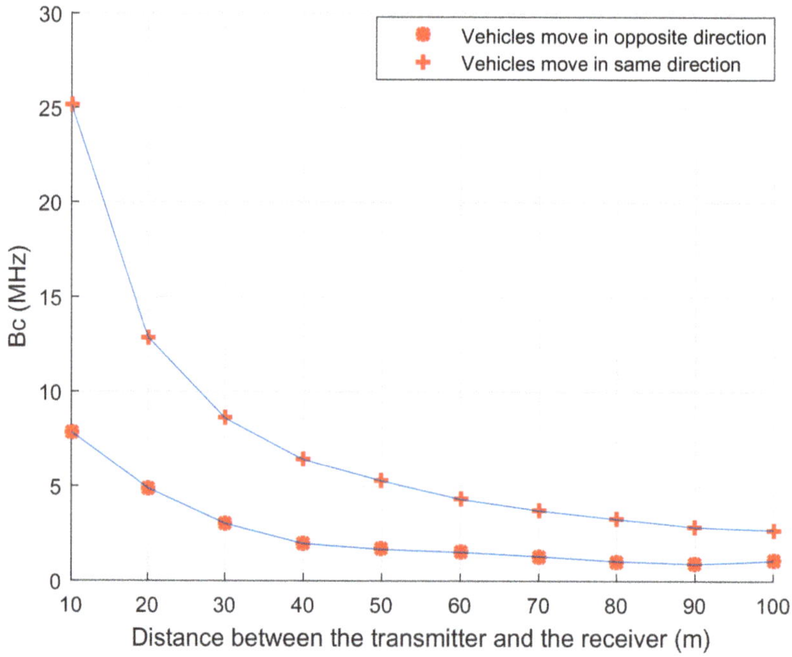

Figure 4.17. Coherence bandwidth for one reflection system with 0.9 correlation level when vehicles move in the same and opposite directions.

Low RMS delay spread values lead to higher coherence bandwidth because in the same direction movement, the NLoS path can be considered dominant. Also, this result shows the importance of the scattering environment of the channel, as the paths take less time to reach the receiver, so the coherence bandwidth of the channel increases.

4.7 Chapter summary

In this chapter, we model the V2V-VLC communication system as 2 × 2 MIMO, considering the real-time environment, practical reflectors, and real ambient noise. Then, the channel of communication is modeled considering LoS and NLoS paths based on upcoming IEEE 802.15.7r1. We derive the CIR expressions for LoS and NLoS paths. Finally, we calculate the channel parameters such as BER, SNR, RMS delay spread, coherence bandwidth, and coherence time when the vehicles move in the same direction and when they move in the opposite direction.

References

[1] Light emiting diode available online: http://www.powerbulbs.com (Accessed: July 1, 2023)
[2] Photo diode parameters http://www.hamamatsu.com (Accessed: June 28, 2023)
[3] Araki N and Yashima H 2005 A channel model of optical wireless communications during rainfall *2nd Int. Symp. on Wireless Communication Systems, 2005* 205–9

[4] Carruthers J B and Kahn J M 1996 Multiple-subcarrier modulation for nondirected wireless infrared communication *IEEE J. Sel. Areas Commun.* **14** 538–46
[5] Carruthers J B and Kahn J M 1997 Modeling of nondirected wireless infrared channels *IEEE Trans. Commun.* **45** 1260–8
[6] Carruthers S M and Carroll J B 2005 Statistical impulse response models for indoor optical wireless channels *Int. J. Commun. Syst.* **18** 1260–8
[7] Li C, Yi Y, Lee K and Lee K 2014 Performance analysis of visible light communication using the STBC-OFDM technique for intelligent transportation systems *Int. J. Electron.* **101** 1117–33
[8] Chen A L, Wu H P, Wei Y L and Tsai H M 2016 Time variation in vehicle-to-vehicle visible light communication channels *2016 IEEE Vehicular Networking Conf. (VNC)* 1–8
[9] Cheng L, Henty B E, Stancil D D, Bai F and Mudalige P 2007 Mobile vehicle-to-vehicle narrow-band channel measurement and characterization of the 5.9 GHz dedicated short range communication DSRC frequency band *IEEE J. Sel. Areas Commun.* **25** 1501–16
[10] Cui Z, Wang C and Tsai H M 2014 Characterizing channel fading in vehicular visible light communications with video data *Vehicular Networking Conf. (VNC), 2014 IEEE* 226–9
[11] Do T-H and Yoo M 2012 Received power and SNR optimization for visible light communication system *Fourth International Conf. on Ubiquitous and Future Networks (ICUFN)* 151 6–7
[12] Luo P, Ghassemlooy Z, Minh H L, Bentley E, Burton A and Tang X 2015 Performance analysis of a car-to-car visible light communication system *Appl. Opt.* **54** 1696–706
[13] Farahneh H, Hussain F, Hussain R and Fernando X 2018 A novel optimal precoder and equalizer in 2 × 2 MIMO VLC systems for vehicular application *2018 IEEE Globecom Workshops (GC Wkshps)* 1–6
[14] Farahneh H, Khalifeh A and Fernando X 2016 An outdoor multi path channel model for vehicular visible light communication systems *2016 Photonics North (PN)* 1
[15] Pan J, Chen Q, Qiang W and Geng L 2015 Results of a new polarimetric BRDF simulation of metallic surfaces *Infrared Phys. Technol.* **72** 58–67
[16] Hayasaka N and Ito T 2007 Channel modeling of nondirected wireless infrared indoor diffuse link *Electron. Commun. Japan (Part I: Commun.)* **90** 9–19
[17] Kahn J M and Barry J R 1997 Wireless infrared communications *Proc. IEEE* **85** 265–98
[18] Kahn J M, Krause W J and Carruthers J B 1995 Experimental characterization of non-directed indoor infrared channels *IEEE Trans. Commun.* **43** 1613–23
[19] Kinoshita M, Yamazato T, Okada H, Fujii T, Arai S, Yendo T and Kamakura K 2014 Motion modeling of mobile transmitter for image sensor based I2V-VLC *V2I-VLC, and V2V-VLC In Globecom Workshops (GC Wkshps)* 450–5
[20] Lee I E, Sim M L and Kung F W L 2011 A dual-receiving visible-light communication system under time-variant non-clear sky channel for intelligent transportation system *2011 16th European Conf. on Networks and Optical Communications (NOC)* 153–6
[21] Lee S J, Kwon J K, Jung S Y and Kwon Y H 2012 Simulation modeling of visible light communication channel for automotive applications *2012 15th Int. IEEE Conf. on Intelligent Transportation Systems (ITSC)* 463–8
[22] Luo P, Ghassemlooy Z, Le Minh H, Bentley E, Burton A and Tang X 2014 Fundamental analysis of a car to car visible light communication system *2014 9th Int. Symp. on Communication Systems, Networks Digital Signal Processing (CSNDSP)* 1011–6

[23] Kojima S, Sivak M, Flannagan M J and Traube E C 2005 A market weighted description of low-beam headlighting patterns in the US *Proc. IEEE* **1** 61–71

[24] Mecklenbrauker C F, Molisch A F, Karedal J, Tufvesson F, Paier A, Bernado L, Zemen T, Klemp O and Czink N 2011 Vehicular channel characterization and its implications for wireless system design and performance *Proc. IEEE* **99** 1189–212

[25] Miramirkhani F and Uysal M 2015 Channel modeling and characterization for visible light communications *IEEE Photon.* J **7** 1–16

[26] Ministry of Transportation of Canada. Headlamps properties http://www.mto.gov.on.ca/english/dandv/driver/handbook (Accessed: June 10, 2023)

[27] Headlamps properties. Headlamps properties http://en.wikipedia.org/wiki/Headlamp/2014 (Accessed: July 5, 2024)

[28] Sarbazi E, Uysal M, Abdallah M and Qaraqe K 2014 Ray tracing based channel modeling for visible light communications *2014 22nd Signal Processing and Communications Applications Conf. (SIU)* 702–5

[29] Schulze H 2016 Frequency-domain simulation of the indoor wireless optical communication channel *IEEE Trans. Commun.* **64** 2551–62

[30] Jung S-Y, Lee S, Kyun Kwon J and Kwon Y-H 2012 Evaluation of visible light communication channel delay profiles for automotive applications *EURASIP J. wirel. Commun. Netw.* **1** 370–9

[31] Lee S, Kwon J K, Jung S-Y and Kwon Y-H 2012 Evaluation of visible light communication channel delay profiles for automotive applications *EURASIP J. Wirel. Commun. Netw.* **2012** 370

[32] Uysal M, Ghassemlooy Z, Bekkali A, Kadri A and Menouar H 2015 Visible light communication for vehicular need headlamp beam pattern model *IEEE Veh. Technol. Mag.* **10** 45–53

[33] Varela M S and Sanchez M G 2001 RMS delay and coherence bandwidth measurements in indoor radio channels in the UHF band *IEEE Trans. Veh. Technol.* **50** 515–25

[34] Jungnickel V, Pohl V, Nonnig S and von Helmolt C 2002 A physical model of the wireless infrared communication channel *IEEE J. Sel. Areas Commun.* **20** 631–40

[35] Wang C, Yu H Y and Zhu Y J 2016 A long distance underwater visible light communication system with single photon avalanche diode *IEEE Photon.* J **8** 1–11

[36] Popoola W, Ghassemlooy Z and Rajbhandari S 2013 *Optical Wireless Communications System and Channel Modelling with MATLAB* (CRC Press, Taylor and Francis Group)

IOP Publishing

Visible Light Communications (Second Edition)
Vehicular applications
Xavier Fernando and Hasan Farahneh

Chapter 5

Adaptive optical OFDM for V2X communications

VLC channels have destructive effects due to scattering, noise and multipath dispersion as we have seen in previous chapters. This results in inter-symbol interference (ISI) and low signal-to-noise ratio (SNR) that leads to low data rate. It is well known that the orthogonal frequency division multiplexing (OFDM) is very effective to address this issue [15]. However, the basic OFDM has to be appropriately modified into *unipolar versions* such as direct current optical OFDM (DCO-OFDM) and asymmetrically clipped optical OFDM (ACO-OFDM) schemes to suit the visible light communication (VLC) and optical wireless communication (OWC) environments.

This chapter gives a brief introduction to optical OFDM (O-OFDM). The DCO-OFDM and ACO-OFDM systems are compared in terms of power and spectral efficiency. Then the vehicle-to-everything (V2X) communication system with adaptive OFDM modulation is studied considering the multipath channel model we derived in section 4.3.2. The effects of sunlight plus clipping noise are also incorporated. Moreover, the performance of the V2X-VLC system is optimized using singular value decomposition (SVD) technique and adaptive bit-loading. Simulation results show DCO-OFDM yields in higher data rate while ACO-OFDM is more tolerant to LED nonlinearity.

5.1 The OFDM principle

OFDM has been implemented in many wireless radio frequency (RF) and wireline standards and has also proved to be beneficial in VLC systems [13]. O-OFDM schemes have better optical power efficiency, reduced ISI and very low bit error rate (BER), compared to conventional optical modulation schemes such as on–off keying (OOK) and pulse position modulation (PPM) [19].

doi:10.1088/978-0-7503-6049-4ch5

Multiple-input–multiple-output (MIMO) systems have gained considerable attention in VLC networks due to their high data rate capabilities over longer distances despite multipath fading and interference effects. The combination of OFDM and MIMO is considered a powerful physical layer solution for high-speed vehicular VLC systems, especially for accommodating bandwidth-hungry applications [16].

Previously, the combination of MIMO and OFDM has been mostly used for non-adaptive systems or for single-modulation schemes. Few existing works have realized vehicle-to-vehicle (V2V)-VLC systems as MIMO systems or used optical OFDM. Adaptive transmission in VLC has been explored before, but mostly for indoor applications [11]. Researchers have only assumed a Gaussian channel with a fixed modulation scheme, with an aim to increase the data rate.

OFDM is a multicarrier modulation scheme, where the individual modulated subcarriers are transmitted in a multiplexed setup in parallel. Whereas in single-carrier systems the symbol duration t_S is given by the reciprocal baud rate $1/R$, in multicarrier systems it is N_{SC}/R, where N_{SC} is the number of parallel transmitted subcarriers. If the subcarriers are shifted close together in frequency domain to achieve the highest spectral efficiency, the independence or orthogonality of the subcarriers is required. Then all subcarriers may overlap, but, at the frequency of the nth subcarrier, the power of all other subcarriers is zero, as shown in figure 5.1. This orthogonality is achieved if the subcarriers frequencies are equidistant in frequency domain leading to the frequency of the kth subcarrier.

$$f_k = f_0 + k\, f_s \tag{5.1}$$

where f_0 is the lowest frequency of the OFDM spectra, f_s is the frequency spacing, and k is the subcarriers index with $k = 0, \ldots, N_{SC}$. The superposition of the independent modulated subcarriers is typically performed by the inverse discrete Fourier transform (IDFT), where the input channels are spaced equivalently according to (5.1). For modulation of subcarriers, phase modulation is primarily applied with four levels leading to a QPSK or higher level combined amplitude and phase modulation with QAM 8 up to QAM 64. For demodulation of the orthogonal subcarriers, a matched filter bank is required. The DFT can replace the analog demodulator realization exhibiting desired functionality. In practice, the DFT and IDFT are realized by the fast Fourier transform (FFT) algorithm and its inverse, IFFT, exhibiting an effective realization for computation of complex values in a digital domain. Therefore, the number of sampling points in the time domain, as well as the number of grid points in the frequency domain, are a power of 2.

Cyclic prefix

The basic concept behind the OFDM cyclic prefix is quite straightforward. The cyclic prefix performs two main functions. The cyclic prefix provides a guard interval to eliminate ISI from the previous symbol. It repeats the end of the symbol so the linear convolution of a frequency-selective multipath channel can be modeled as circular convolution, which in turn may transform to the frequency domain via a DFT.

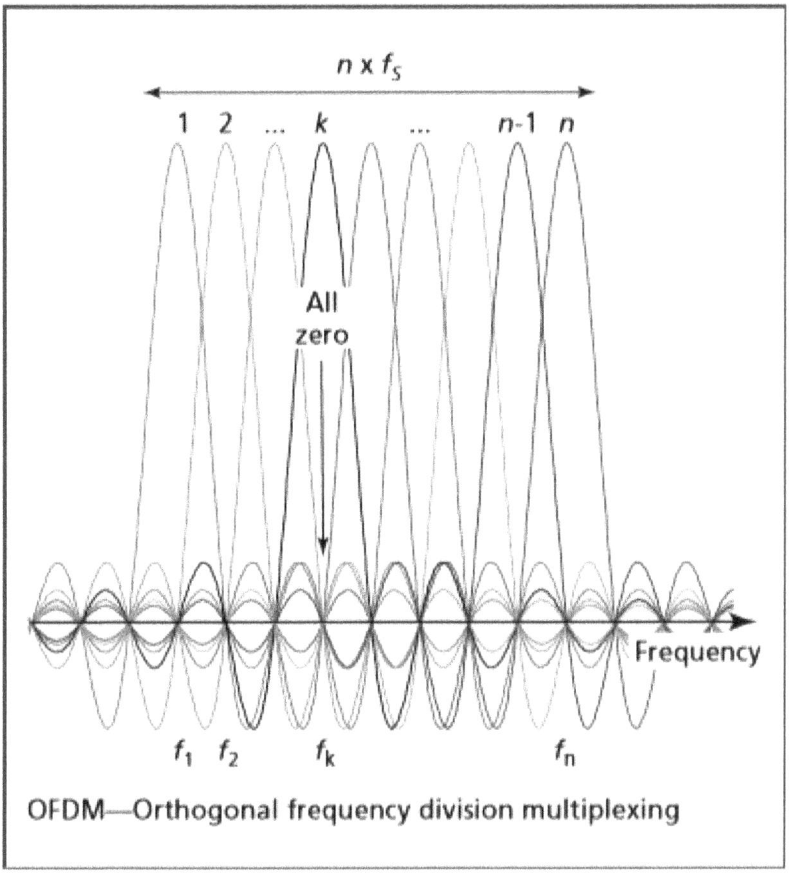

Figure 5.1. Spectrum of an OFDM signal with overlapping subcarriers.

This approach accommodates simple frequency domain processing, such as channel estimation and equalization.

The cyclic prefix is created so that each OFDM symbol is preceded by a copy of the end part of that same symbol. Different OFDM cyclic prefix lengths are available in various systems. For example within long term evolution (LTE) a normal length and an extended length are available and after Release 8 a third extended length is also included, although not normally used.

There are several advantages and disadvantages attached to the use for the cyclic prefix within OFDM. The addition of the cyclic prefix adds robustness to the OFDM signal. The guard interval introduced by the cyclic prefix enables the effects of ISI to be reduced. However, the cyclic prefix reduces the overall data rate as it re-transmits data that is already being transmitted.

Overall, the use of a cyclic prefix is standard within OFDM and it enables the performance to be maintained even under conditions when levels of reflections and multipath propagation are high (figure 5.2).

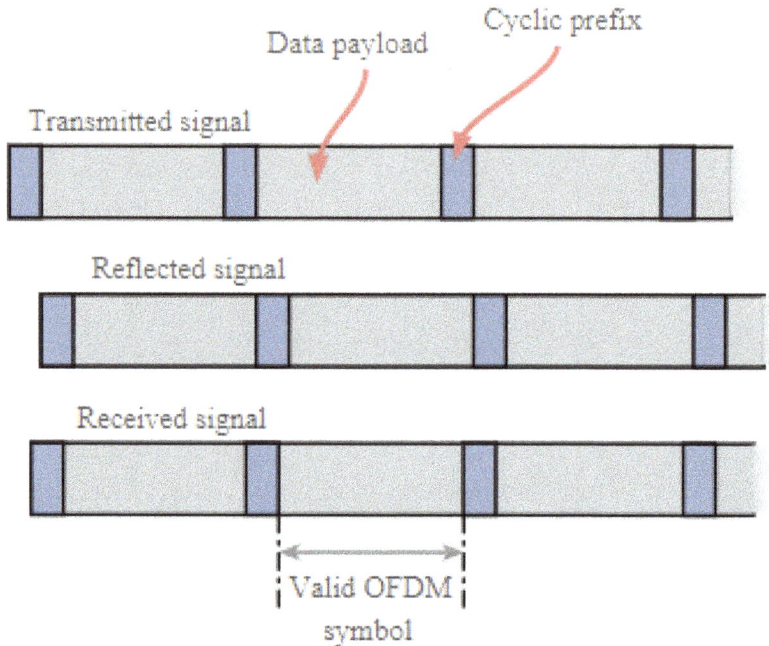

Figure 5.2. OFDM cyclic prefix.

5.2 Optical (unipolar) OFDM

In general, electrical OFDM signals are complex valued data signals. To convert a complex valued data signal into the optical domain, the signal can be electrically up-converted using an electrical intermediate frequency (IF) carrier to modulate the signal upon applying a complex electrical I/Q mixer. The resulting up-converted electrical OFDM data signal can be electro-optic (e/o) converted using a state-of-the-art amplitude modulator. Alternatively, a complex electro-optic in-phase/quadrature (I/Q) modulator can be driven directly by the complex electrical OFDM signal and thus converted into the optical domain. The transmitter setups are summarized in figure 5.3 [3].

In optical receivers, typically, simple photodiodes are applied which operate according to the square law detection scheme. Figure 5.4 illustrates O-OFDM receiver setups. It is impossible to transfer the optical OFDM signal directly into the electrical domain. Instead of direct detection of the pure optical OFDM signal, an optical carrier must be delivered either by the transmitter (using direct detection OFDM) or by a local oscillator in the receiver (using coherent optical OFDM) via a heterodyne or intradyne approach, shown, respectively, in figure 5.4. In the heterodyne scenario, the optical OFDM signal is converted into a real valued electrical OFDM signal at an intermediate frequency (IF). Using a subsequent electrical I/Q demodulator, the real and imaginary components are available in the baseband. In the intradyne approach, the local oscillator wavelength is close to the transmitter wavelength. The OFDM signal then beats with the local oscillator (LO)

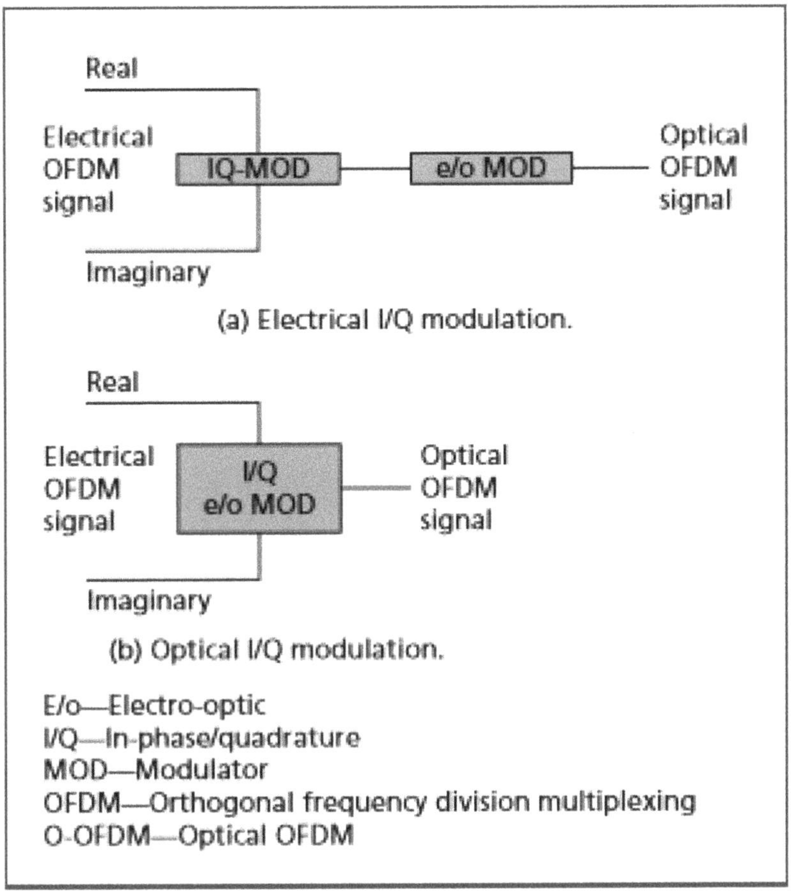

Figure 5.3. O-OFDM transmitter setups.

signal in an optical 90° hybrid to obtain the I and Q components of the signal. To realize a 10 Gb s^{-1} optical OFDM system, analog–digital interfaces with a rate of 10 giga samples per second (GSa s^{-1}) may be applied.

5.3 Related work in adaptive modulation for VLC

In literature, there has been limited research into O-OFDM-V2V-VLC systems. In this section, some of these works will be reviewed. Performance of a VLC system using space-time block coding (STBC)-OFDM technique for ITS was investigated in [5]. They modeled the outdoor optical channel as a Rician channel. They proposed STBC-OFDM coding to reduce the influence of multipath dispersion associated with the channel to achieve a BER of 10^{-6} even at a low SNR. The authors in [10] presented a single-input–single-output (SISO)-V2V-VLC system with DCO-OFDM. They proposed a new automotive VLC system based on the optical communication image sensor (OCI) as a receiver. Then, they applied DCO-OFDM to achieve more flexible and effective signal transmission. Also, they considered the characteristics of

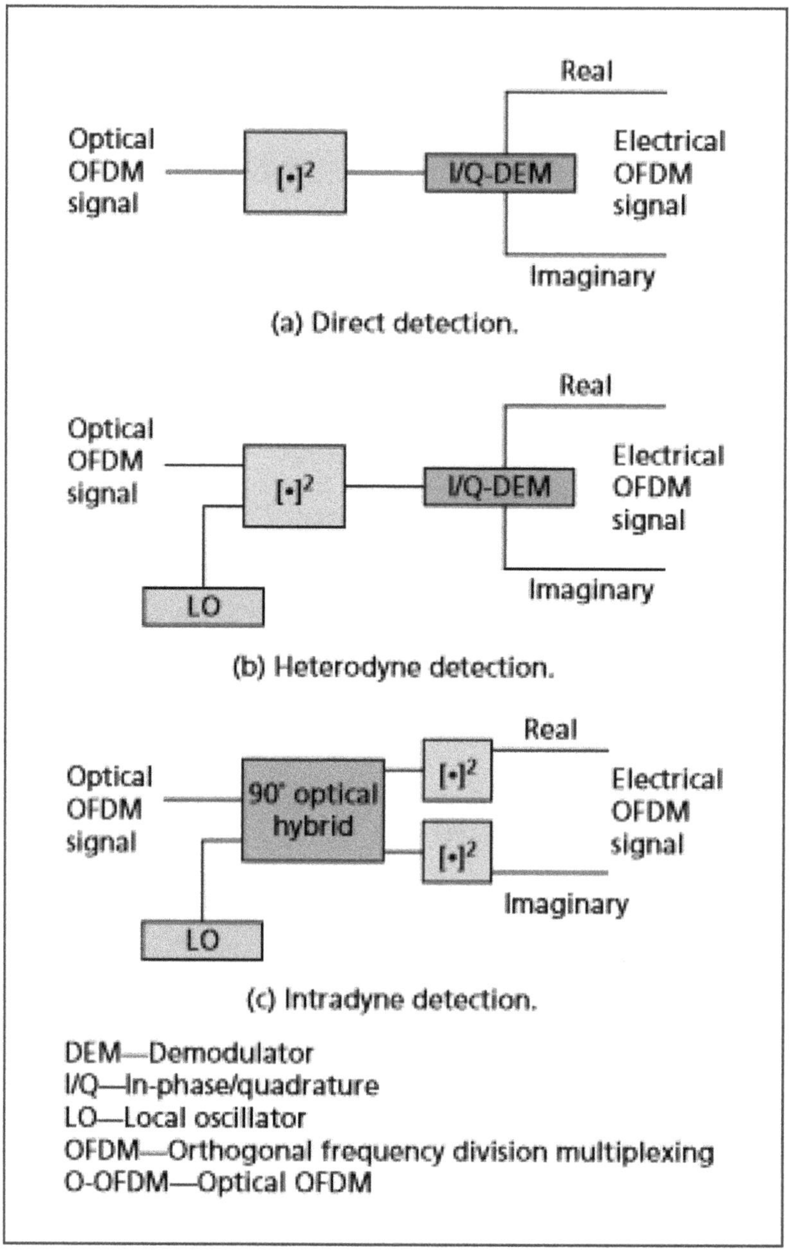

Figure 5.4. Optical-OFDM receiver setups.

the LED and OCI to select the appropriate parameter values for signal processing to achieve higher data rates. The authors in [4] introduced the concept of an environment-adaptive VLC receiver for a vehicular communication system in a visible light environment to mitigate the effects of problematic conditions such as supporting long-distance communications in dynamic traffic situations and in unfriendly atmospheric conditions. In [14], the authors suggested a DCO-OFDM-based MIMO transmission scheme for vehicular communication. They evaluated the performances of different MIMO modes such as spatial multiplexing, and repetition code with different modulation techniques and different transmitter–receiver selection. They found the spatial multiplexing technique suffers from channel correlation, while repetition code is unsuitable for high modulation orders.

5.4 Optical OFDM scheme of a V2V-VLC system

This section outlines the investigation scheme for V2V-VLC channel and analyzes it with various types of OFDM techniques. In this study, the model presented in section 4.3.2 and shown in figure 4.3 is modified to include O-OFDM with an adaptive modulation scheme. This reduces ISI and increases the data rate. Figure 5.5 shows the block diagram of the proposed O-OFDM scheme for a V2V-VLC system (table 5.1).

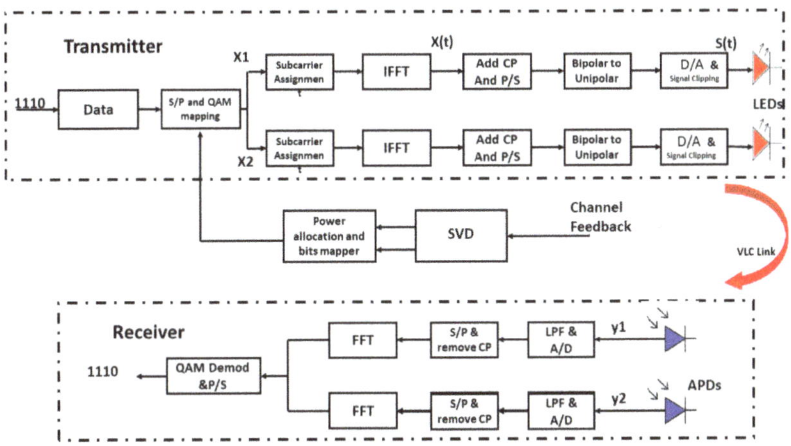

Figure 5.5. Simplified diagram of the V2V-VLC-OFDM scheme.

Table 5.1. Brief comparison of DCO-OFDM and ACO-OFDM.

Feature	DCO-OFDM	ACO-OFDM
DC Bias	Required	Not required
Energy efficiency	Lower (due to DC bias)	Higher
Complexity	Lower	Higher (complex receiver needed)
Spectral efficiency	Lower	Higher (half the bandwidth needed)
Handling of LED nonlinearities	Poor (affected by DC bias)	Better (No DC component)
Clipping noise	Usually no	High probability
Power consumption	Higher	Lower

In figure 5.5, a serial-to-parallel (S/P) converter is used to split a binary source of data into two data streams $\mathbf{x}_1$, $\mathbf{x}_2$. Both streams are modulated using QAM and the resulting symbols are assigned to subcarriers. Afterward, IFFT is performed and data is converted back to a single stream using a parallel-to-serial (P/S) converter after adding a cyclic prefix (CP). Then, the signal is changed to unipolar using one of the O-OFDM schemes. Finally, the data is converted to analog signal $s(t)$ using a digital-to-analog (D/A) converter along with low-pass filter (shaping filter).

5.4.1 DCO-OFDM modulation scheme

DCO-OFDM is a form of O-OFDM with a DC bias added. In such systems, the frequency domain signal is transformed into the time domain by using the IDFT, which can be implemented by using an IFFT. In this study, the length of the IFFT is assumed to be N. The modulated signal in the frequency domain must be conjugate symmetric to ensure that the time domain signal is real. Moreover, a DC bias is added to guarantee that the transmitted signals are nonnegative.

In the frequency domain, the modulated signals satisfy the following conditions

$$\begin{cases} X(0) = X\left(\frac{N}{2}\right) = 0 \\ X(K) = X^*(N-K) \end{cases} \qquad K = 1, 2, \dots, N-1 \tag{5.2}$$

where N is the size of IFFT, and * is conjugate sign.

Thus, the input vector to the IFFT block, in figure 5.5, $\mathbf{x}$, can be given as

$$\mathbf{x} = [0 \quad \mathbf{s}_1 \quad \mathbf{s}_2 \quad \cdots \quad \mathbf{s}_{N/2-1} \quad 0 \quad \mathbf{s}^*_{N/2-1} \quad \cdots \quad \mathbf{s}^*_2 \quad \mathbf{s}^*_1] \tag{5.3}$$

The time domain signal can be obtained by performing IFTT and it can be expressed as

$$\begin{aligned} x[n] &= \frac{1}{N}\sum_{K=0}^{N-1} X(K)\, e^{\frac{j2\pi Kn}{N}} \\ &= \frac{1}{N}\sum_{K=0}^{\frac{N}{2}-1} X(K)\, e^{\frac{j2\pi Kn}{N}} + \frac{1}{N}\sum_{\frac{N}{2}+1}^{N-1} X(K)\, e^{\frac{j2\pi Kn}{N}} \\ &= \frac{1}{N}\sum_{K=1}^{\frac{N}{2}-1}\left(X(K)\, e^{\frac{j2\pi Kn}{N}} + X^*(K)\, e^{\frac{-j2\pi Kn}{N}}\right) \\ &= \frac{2}{N}\sum_{K=1}^{\frac{N}{2}-1}\left(a(K)\cos\left(\frac{2\pi Kn}{N}\right) - b(K)\sin\left(\frac{2\pi Kn}{N}\right)\right) \end{aligned} \tag{5.4}$$

It is assumed that $X(K) = a(K) + jb(k)$ is a zero mean complex random variable with variance σ_K^2, $X(K)$, $K = 1, \dots, N/2 - 1$, are independent, and $a(K)$ and $b(K)$ are independent real random variables with a zero mean and variance $\sigma_K^2/2$.

According to the central limiting theorem, $x[n]$ is a Gaussian random variable with zero mean, that is, $E[x(n)] = 0$, where $E[.]$ stands for expectation.

The variance of $x[n]$ can be derived as

$$\begin{aligned}\sigma_x^2 &= E[x[n]\, x^*[n]] \\ &= \frac{4}{N^2}\sum_{K=1}^{\frac{N}{2}-1}\left(\frac{\sigma_K^2}{2}\cos^2\left(\frac{2\pi Kn}{N}\right)+\frac{\sigma_K^2}{2}\sin^2\left(\frac{2\pi Kn}{N}\right)\right) \\ &= \frac{2}{N^2}\sum_{k=1}^{N/2-1}\sigma_k^2\end{aligned} \tag{5.5}$$

In DCO-OFDM technique, data is assigned to both odd and even subcarriers. A real signal can be then obtained by imposing the Hermitian symmetry property as shown in (5.2). This property implies that half of the OFDM subcarriers are utilized to generate the real-time domain signal [9].

The output of the D/A converter is an analog real-time signal, which is shown in figure 5.6(a). This signal cannot be used to modulate the LEDs because of its bipolar

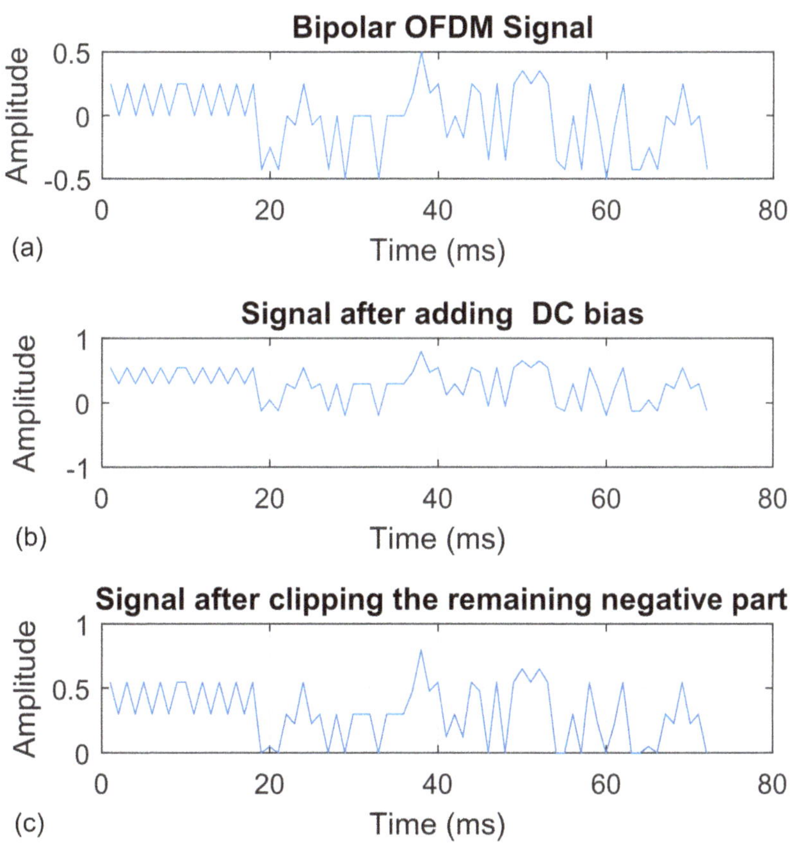

Figure 5.6. (a) OFDM bipolar signal. (b) The signal after adding DC bias voltage. (c) The signal after clipping the remaining negative.

nature. A certain bias value is added to this signal to convert it to unipolar and is plotted in figure 5.6(b). Afterward, any remaining negative values are clipped at zero and are plotted in figure 5.6(c), which introduces distortion noise. The added DC value increases the transmitter power requirement, while the clipping introduces clipping noise in both the even and odd subcarriers.

The transmitted signal $s(t)$ can be expressed as

$$s(t) = x(t) + b_{dc} - e_c(t) \tag{5.6}$$

where b_{dc} is the bias voltage, and $e_c(t)$ is clipping noise with a variance $\sigma^2_{c_{DCO}}$.

To minimize the clipping noise, the bias voltage b_{dc} is given by [20]

$$b_{dc} = 2\,\sigma_x\,\zeta \tag{5.7}$$

where ζ is the biasing ratio and is given by [20]

$$\zeta = \frac{b_{dc}}{c_u - c_l} = \frac{-c_l}{c_u - c_l} \tag{5.8}$$

where c_u and c_l denote the upper and the lower clipping level, respectively.

The variance of the clipping noise $\sigma^2_{c_{DCO}}$ is given by [18]

$$\begin{aligned}\sigma^2_{c_{DCO}} = \sigma_x^2 + (b_{dc}^2 - \sigma_x^2)Q\left(\frac{b_{dc}}{\sigma_x}\right) - b_{dc}\sigma_x g\left(\frac{b_{dc}}{\sigma_x}\right)\\ - \sigma_x^2 Q\left(\frac{-b_{dc}}{\sigma_x}\right)^2 - \left(\sigma_x g\left(\frac{b_{dc}}{\sigma_x}\right) - b_{dc} Q\left(\frac{b_{dc}}{\sigma_x}\right)\right)^2\end{aligned} \tag{5.9}$$

where $Q(.)$ is the tail probability of the standard Gaussian distribution, $g(.)$ is the PDF of the clipping noise which follows Gaussian distribution and is given by [7]

$$g(u) = \frac{1}{\sqrt{2\pi\sigma_x^2}}\, e^{-\frac{\left(u - \frac{\zeta}{2}\sigma_x\right)^2}{2\sigma_x^2}} \tag{5.10}$$

The average transmitted optical power is expressed as

$$E[s(t)] = \sigma_x\left(\frac{\zeta}{2} + \frac{1}{\sqrt{2\pi}} e^{\frac{-\zeta^2}{8}} - \frac{\zeta}{2} Q\left(\frac{\zeta}{2}\right)\right) \tag{5.11}$$

For fixed modulation scheme, the maximum channel capacity (bps/Hz) of the DCO-OFDM-VLC system can be given as

$$R_{DCO} = \frac{1}{N + N_{cp}} \sum_{k=1}^{N/2-1} \log_2\left(1 + \frac{|H(k)|^2\,\sigma_k^2}{\sigma_n^2 + |H(k)|^2 \sigma^2_{c_{DCO}} + \sigma^2_{ISI}}\right) \tag{5.12}$$

where $|H(k)|$ indicates the kth channel transfer function, N_{cp} indicates the number of cycling prefix of the OFDM, σ_n^2 indicates the ambient noise power which is given by (2.7), and $\sigma_{\rm ISI}^2$ is ISI noise and is given by

$$\sigma_{\rm ISI}^2 = \gamma^2\left(\sum_{k=1}^{N}\int_{T_d}^{\infty} h_k(t) \otimes s(t)\, d(t)\right)^2 \tag{5.13}$$

where T_d denotes the transmitted optical pulse duration, and k is channel index.

The maximum channel capacity is subject to (optical power constraint) $0 \leqslant E[s(t)] \leqslant P_t$, where P_t is the transmitted power.

5.4.2 ACO-OFDM modulation scheme

In the ACO-OFDM technique, the modulated data is assigned to the odd indexed subcarriers, while the even indexed subcarriers are set to zero. Also, signals in this scheme can be transmitted without any DC bias.

Referring to figure 5.5, the input vector $\mathbf{x}$ of the IFFT block can be given as

$$\mathbf{x} = [0 \quad \mathbf{s}_0 \quad 0 \quad \mathbf{s}_1 \quad \cdots \quad \mathbf{s}_{\mathbf{N}/4-1} \; 0 \; \mathbf{s}_{\mathbf{N}/4-1}^* \quad \cdots \quad 0 \quad \mathbf{s}_1^* \; 0 \; \mathbf{s}_0^*] \tag{5.14}$$

After performing IFFT, the signal can be written as

$$x\left[n + \frac{N}{2}\right] = -x[n] \qquad n = 0, 1, \ldots, \frac{N}{2} - 1 \tag{5.15}$$

Also, $x[n]$ can be considered a Gaussian random variable with zero mean and variance is given by

$$\sigma_x^2 = \frac{2}{N^2}\sum_{k=1}^{N/4}\sigma_{2k-1}^2 \tag{5.16}$$

In the ACO-OFDM scheme, bipolar to unipolar conversion is achieved by clipping the negative amplitudes which will not introduce any distortion to the transmission. LED turn on voltage V_{tov}, is added to shift the unipolar signal to the dynamic range of the LED. However, any amplitude levels exceeding the upper limit of the LED dynamic range $V_{\max}$ must be clipped, thus resulting in clipping noise.

The transmitted signal in an ACO-OFDM is written as

$$s(t) = \begin{cases} x(t) & x(t) > 0 \\ 0 & x(t) \leqslant 0 \end{cases} \tag{5.17}$$

where $s(t)$ is a clipped Gaussian signal.

The PDF of $s(t)$ is written as [11]

$$f_{s(t)}(u) = \begin{cases} 0.5 & u = 0 \\ \dfrac{1}{\sqrt{2\pi\sigma_x^2}} e^{\frac{-u^2}{2\sigma_x^2}} & u > 0 \end{cases} \tag{5.18}$$

The average power of $s(t)$ can be calculated as

$$E[s(t)] = 0.5 \times 0 + \int_0^{\infty} u \frac{1}{\sqrt{2\pi\sigma_x^2}} e^{\frac{-u^2}{2\sigma_x^2}} du = \frac{\sigma_x}{\sqrt{2\pi}} \tag{5.19}$$

The channel capacity of fixed modulation scheme for the ACO-OFDM is given by

$$R_{\text{ACO}} = \frac{1}{N + N_{cp}} \sum_{k=1}^{N/4} \log_2\left(1 + \frac{|H(k)|^2 \sigma_{2k-1}^2}{4\sigma_n^2 + |H(k)|^2 \sigma_{c_{\text{ACO}}}^2 + \sigma_{ISI}^2}\right) \tag{5.20}$$

where $\sigma_{c_{\text{ACO}}}^2$ is the variance of the clipping noise and is given by [7]

$$\sigma_{c_{\text{ACO}}}^2 = \frac{1}{2}\sigma_x^2 \, Q(\sqrt{2}\eta)(1 + \eta^2) - \frac{\sigma_x^2 \eta}{2\sqrt{\pi}} e^{-\eta^2} \tag{5.21}$$

where ($\eta = \frac{I_{\max} - I_{\text{bias}}}{\sigma_x}$) is a dimming level, $I_{\max}$ and I_{bias} are the maximum input current and bias current of the LED, respectively. Equation (5.20) shows that the spectral efficiency of ACO-OFDM is reduced by half as compared to DCO-OFDM, since only the odd subcarriers are used in transmission.

In this scheme, also, the maximum channel capacity is subject to (optical power constraint) $0 \leqslant E[s(t)] \leqslant P_t$.

5.5 Performance analysis and bit-loading algorithm

In this section, we present a performance analysis of the V2V-VLC system using bit-loading algorithm. In this model, we use DCO-OFDM and ACO-OFDM schemes for both LoS and NLoS signal paths. An APD at the receiver converts optical signal to electrical signal with an addition noise. The parameter values of the APD can be found in table 2.4 and in [2].

5.5.1 SNR estimation of subcarriers using singular value decomposition (SVD)

The matched filter $p(t)$ is applied to the received signal $y_r(t)$ and then its output is sampled at the rate of T_s, $y_r[n] = y(nT_s)$. The discrete signal can be written as

$$y_r[n] = \sum_{s=1}^{2} (\gamma \ (s_s[n] + e_c[n]) \otimes h_{rs}[n]) + n_r[n]) \tag{5.22}$$

where s, r are the transmitter and the receiver, respectively, with values (1, 2).

The discrete signal in frequency domain can be written as

$$Y_r(k) = \sum_{s=1}^{2} (\gamma \ H_{rs}(k) \, (S_s(k) + E_c(k)) + N_r(k)) \tag{5.23}$$

The estimated channel matrix of the kth subcarrier can be expressed as

$$\mathbf{H}(k) = \begin{bmatrix} H_{11}(k) & H_{12}(k) \\ H_{21}(k) & H_{22}(k) \end{bmatrix} \tag{5.24}$$

5.5.1.1 SVD based precoding design

As RF and VLC have completely different system setups, the optimal structure of the user precoding matrix is also different. SVD-based structure is optimal for RF communication systems [12], but is not optimal for VLC communication, especially when a large number of LEDs are used with a requirement of a large biasing voltage. Since we have assumed a small number of LEDs with small bias requirements in this study, SVD technique can be employed to calculate the gain of each subcarrier. Also, this structure is practical to use in many respects; it can distribute the MIMO channel into a set of parallel non-interference channels [17]. Moreover, an SVD-precoder scheme solves the channel correlation problem which causes low spatial multiplexing gain in visible light MIMO communication.

For the SVD-precoder, the precoding matrix is given as

$$\mathbf{H} = \mathbf{U}\,\mathbf{D}\,\mathbf{V}^* \tag{5.25}$$

where * denotes Hermitian transpose, $\mathbf{V}$ is 2×2 unitary matrix, used in the transmitter to pre-process transmitted signal, $\mathbf{U}$ is a 2×2 receive combining unitary matrix, and $\mathbf{D}$ is a 2×2 diagonal matrix with nonnegative diagonal elements (singular values of matrix $\mathbf{H}$).

Therefore, the channels are decomposed into two independent parallel subchannels. The equivalent channel gain of the two independent subchannels is given as

$$\mathbf{G}_o = \begin{bmatrix} \lambda_{11} & \lambda_{12} & \cdots & \lambda_{1k} \\ \lambda_{21} & \lambda_{22} & \cdots & \lambda_{2k} \end{bmatrix} \tag{5.26}$$

The gain of each subcarrier can be found from the matrix $\mathbf{G}_o$. Therefore, SNR[k] of kth subcarrier can be calculated as

$$\mathrm{SNR}[k] = \frac{\lambda_k}{\sigma_n^2(k)} \tag{5.27}$$

By considering a targeted BER (BER[k]) of the kth subcarrier, the modulation order M_k can be found using (5.28) [13].

$$\mathrm{BER}[k] = \frac{(\sqrt{M_k} - 1)}{\sqrt{M_k}\,\log_2(\sqrt{M_k})}\, erfc\left(\sqrt{\frac{3\,\mathrm{SNR}[k]}{2(M_k - 1)}}\right) \tag{5.28}$$

5.5.2 Adaptive transmission and bit-loading scheme

Bit-loading is performed with respect to the frequency-selective channel and specific modulation order to improve the system's performance. Therefore, subcarriers having high attenuation are re-modulated with lower modulation orders. Similarly, the subcarrier having low attenuation are re-modulated with higher orders. The subcarriers of the O-OFDM signal are modulated by $M - QAM$; the constellation order M can differ from other subcarriers on the basis of SNR. We use table 5.3 to assign the bits for each subcarrier.

The number of bits per symbol of each subcarrier is between 1 and 8. By obtaining M_k for each subcarrier, the transmission rate (bps) for the DCO-OFDM-VLC can be calculated as

$$R_{\mathrm{DCO_{Adapive}}} = \frac{1}{(N + N_{cp})T_s} \sum_{k=1}^{N/2-1} \mathrm{Log}_2 M_k \tag{5.29}$$

For the ACO-OFDM-VLC, the transmission rate can be given as

$$R_{\mathrm{ACO_{Adapive}}} = \frac{1}{(N + N_{cp})T_s} \sum_{k=1}^{N/4-1} \mathrm{Log}_2 M_k \tag{5.30}$$

5.6 Results and discussions

5.6.1 Simulation environment

In this study, MATLAB software is used to model the V2V-VLC system, and parameters/values used in simulations are summarized in tables 2.4, 4.1, and 5.2. Multipath propagation with one reflection system is considered assuming that all the reflectors surfaces have the same reflective factor. Moreover, summer season with clear sky is considered. Also, vehicles are moving at the same speed in the same direction. The city of Toronto is considered for sunlight intensity calculations, whose location coordinates are given in table 5.2. We do Monte Carlo simulation for 5000 iterations for simulating the V2V-VLC channel [8]. To reduce the impact of the channel causing signal distortion, truncated sinc pulse filter is used for pulse shaping.

Table 5.2. Simulation parameters.

The Parameter	Value
Toronto location	43.6532°N, 79.3832°W
Weather	Clear sky
Number of FFT points	64
Subcarrier modulation scheme	4-QAM, 16-QAM, ⋯, 256-QAM
Signal bandwidth	50 MHz
Cyclic prefix length *CP*	8
Targeted BER	10^{-5}
Biased voltage (V)	8 dB, 10 dB
Vehicles speed (same direction)	30 km h^{-1}

Table 5.3. Look-up-table.

	Number of bits/Subcarrier	
M_k	DCO-OFDM-SNR (dB)	ACO-OFDM-SNR (dB)
0	<7	<3
1	7	3
2	8	4
3	9	5
4	10	6
5	11	7
6	12	8
7	⩾14	10
8	NA	⩾12

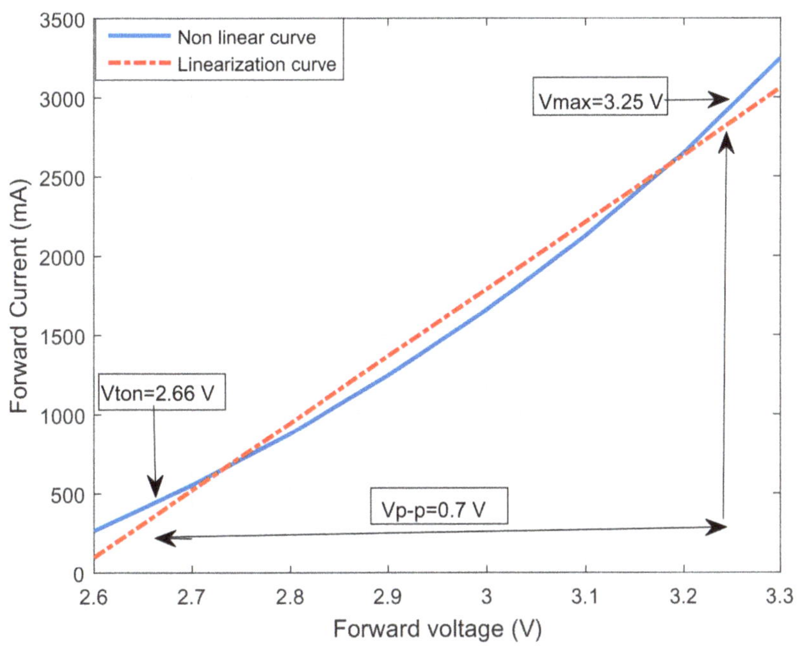

Figure 5.7. *I–V* characteristic curve of the LED.

Also, we use a Philips Ultinon LED 12985BWX2 as transmitter [1]. Moreover, a Hamamatsu-S8664 APD is considered as a receiver [2]. The scheme proposed in [6] is used to model the nonlinearity of anLED's *I–V* property. Figure 5.7 shows that, simulated *I–V* property fits well with the data obtained from the LED datasheet. Moreover, the linear *I–V* curve is obtained by linear regression of the data.

5.6.2 Results and analysis

The system performance is analyzed by considering various parameters for both types of the O-OFDM schemes. The distributions of bit- and power-loading of the DCO-OFDM for 64 subcarriers at luminance intensity (1000–2500 (cd)) with an 8 dB and 10 dB bias is shown in figure 5.8. 128 QAM (7 bits/symbol) is assigned to subcarriers having an SNR $\geqslant$14 dB, binary phase shift keying (BPSK) (1 bit/symbol) is chosen for the subcarriers having SNR $= 7$ dB (as seen with 8 dB bias) and no bits assigned for SNR < 7 dB. The transmitter can obtain the modulation order M_k to be employed for each subcarrier by using table 5.3. Also, figure 5.8 shows, clipping noise is decreased with the increase in bias. Therefore, for 10 dB bias, as the clipping noise is less, the total noise power will depend more on channel frequency response rather than on 8 dB bias. As a result of 8 dB bias, the bit allocation decreases by two bits over 64 subcarriers. Moreover, figure 5.8 shows the power distribution and modulation order of the subcarriers has decreasing trends at a high-frequency spectrum. The reason for this type of behavior depends upon the low-pass property of the LED. In figure 5.9, bit-loading and power distribution for different subcarriers for the ACO-OFDM adaptive modulation scheme are presented under similar channel conditions as in the DCO-OFDM with an 8 dB bias and 10 dB bias, respectively. In this technique, only the odd subcarriers are used for data transmission. Also, the power distribution and modulation order of the subcarriers are decreasing with the high-frequency spectrum. Moreover, a maximum of 256-QAM (8 bits/symbol) is assigned to subcarriers having an SNR of about 12 dB, whereas the BPSK (1 bit/symbol) format is chosen for the subcarriers having

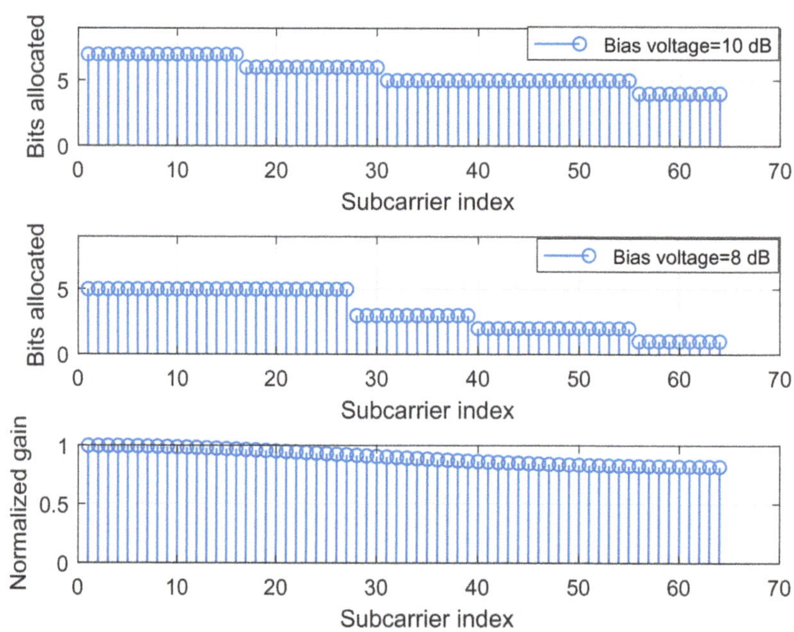

Figure 5.8. Number of loaded bits for DCO-OFDM when $I_T(\alpha_T, \beta_T) = 1500$ (cd).

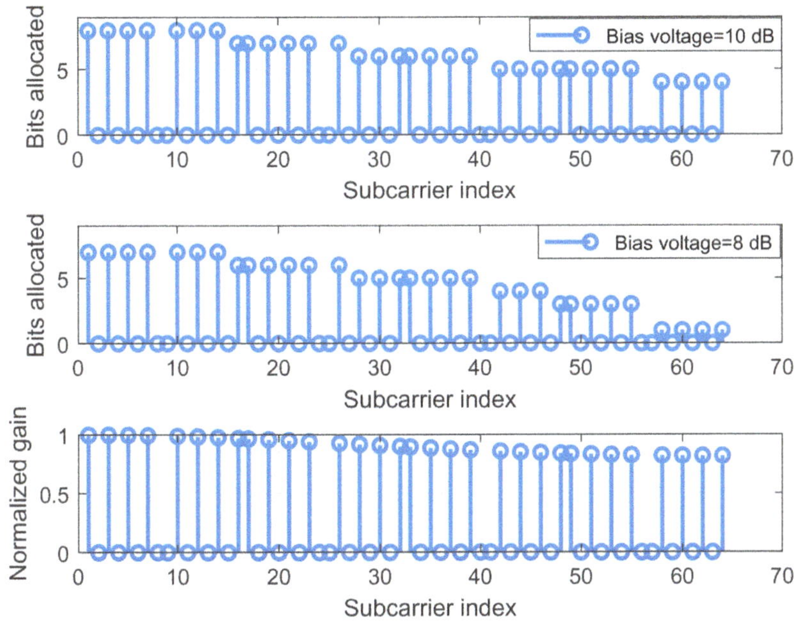

Figure 5.9. Number of loaded bits for ACO-OFDM when $I_T(\alpha_T, \beta_T) = 1500$ (cd).

SNR = 3 dB, and no bits assigned for SNR < 3 dB. Note that the number of assigned bits to the ACO-OFDM scheme is higher than in the DCO-OFDM scheme. This is because the DCO-OFDM experiences higher clipping distortion compared to ACO-OFDM at certain illuminance intensity. Therefore, the number of loaded bits per subcarrier for the DCO-OFDM is generally lower than in the ACO-OFDM. Also, figure 5.9 shows that increasing the bias decreases the clipping noise. As a result of an 8 dB bias, the bit allocation decreases by one bit over 32 subcarriers [8].

5.6.2.1 Transmission rate analysis

In figures 5.10 and 5.11, transmission rates for adaptive DCO-OFDM and ACO-OFDM, are presented. The adaptive transmissions are compared with the fixed modulation schemes while maintaining the desired BER of 10^{-5}. In figure 5.10, it is shown that by employing adaptive transmission in the DCO-OFDM, data rate is significantly improved, compared to fixed modulation. Furthermore, the clipping distortion affects the performance of the DCO-OFDM, above luminance intensity of 1600 (cd). In figure 5.11, the transmission rate for the ACO-OFDM is calculated and it is observed that the clipping distortion degrades the performance when the luminance intensity exceeds 2000 (cd). Moreover, the maximum achievable rate for the ACO-OFDM is lower than the DCO-OFDM for both types of transmissions schemes; i.e. adaptive and fixed. As the ACO-OFDM utilizes a dynamic range more effectively due to half the number of data subcarriers [8].

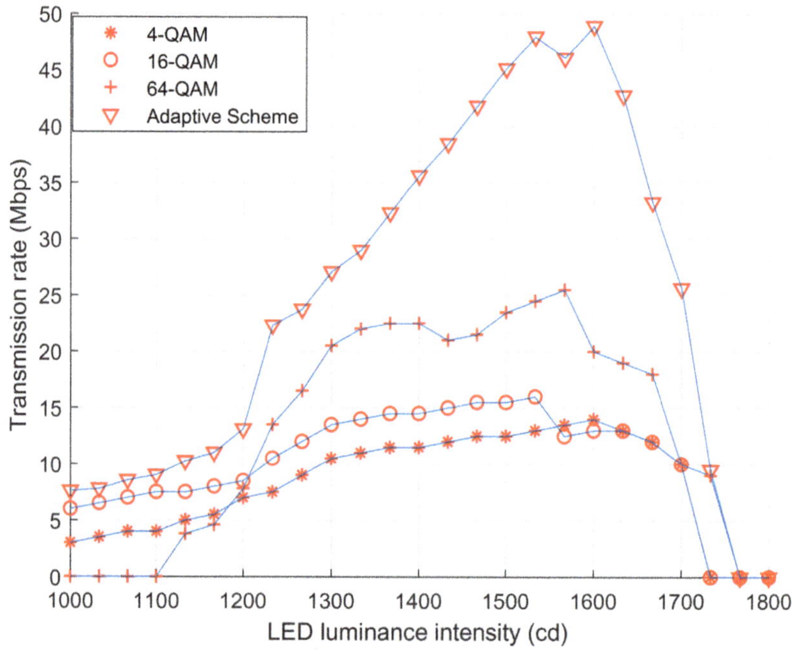

Figure 5.10. The transmission rate of DCO-OFDM with adaptive modulation.

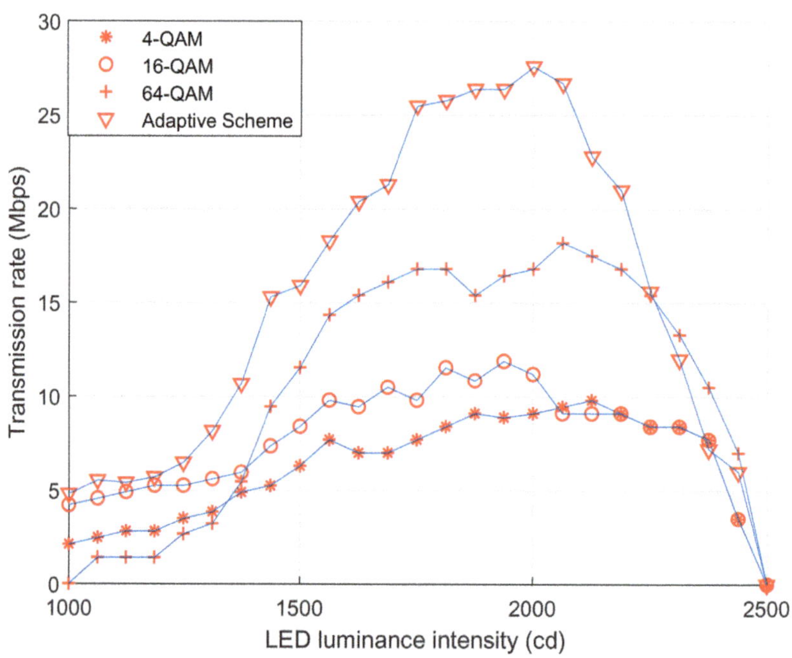

Figure 5.11. The transmission rate of ACO-OFDM with adaptive modulation.

5.6.2.2 BER analysis

BER performance for both the ACO-OFDM and the DCO-OFDM is investigated, for adaptive and fixed modulation schemes. Figures 5.12 and 5.13, show better SNR performance for adaptive modulation than fixed modulation, for both schemes. Moreover, it can be seen that adaptive modulation schemes can meet the target of BER at less SNR value than other QAM schemes. Also, the BER performance is not recorded for SNR range of (0–3 dB) for the ACO-OFDM, and for SNR range of (0–7 dB) for the DCO-OFDM because no bits are transmitted at all in these ranges (according to the bit-loading algorithm). Moreover, the BER performance is not recorded for SNR > 18 dB for both schemes because the clipping distortion degrades the performance. Although, figures 5.12 and 5.13 show a better performance of the ACO-OFDM comparing with the DCO-OFDM. This happens due to the inherent characteristics of the DCO-OFDM, being sensitive to clipping noise [8].

5.7 Chapter summary

This chapter investigates and analyzes adaptive modulation schemes for O-OFDM based V2V-VLC system considering the practical noise environment to fully exploit the transmission resources. Two modulation schemes are used, DCO-OFDM and ACO-OFDM to combat ISI and to improve the data rate. We compare the performance of the adaptive modulation schemes with fixed modulation schemes in terms of BER, and data rate, where the adaptive schemes show superiority in performance.

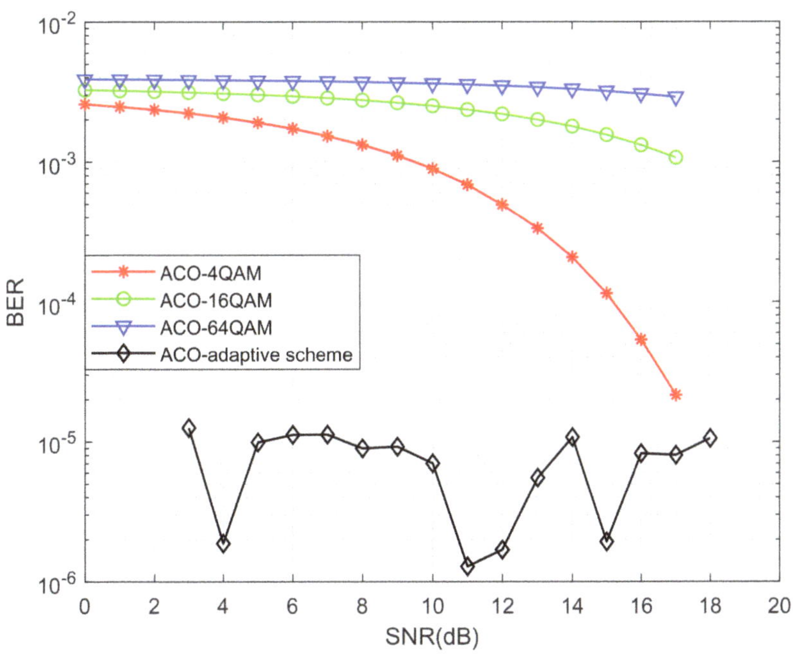

Figure 5.12. SNR versus BER for ACO-OFDM with FFT size = 64.

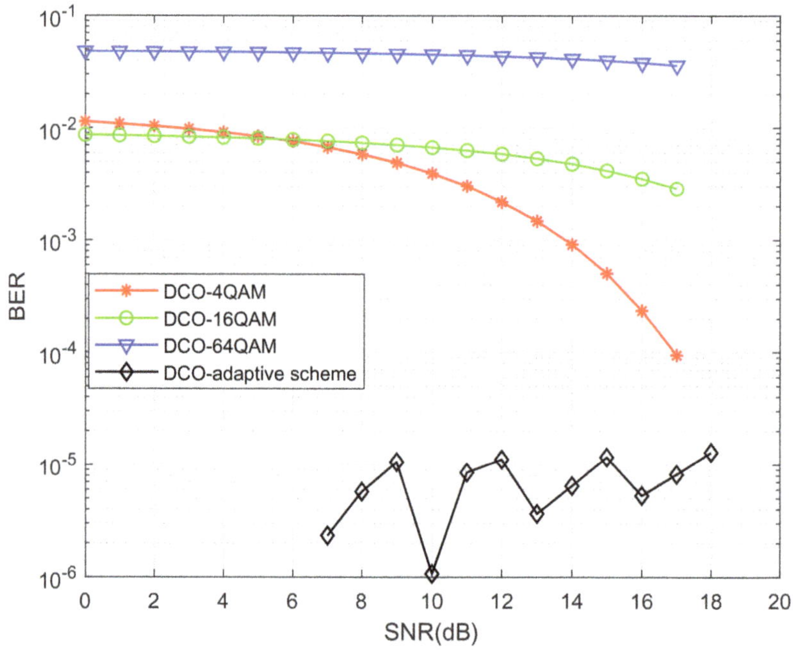

Figure 5.13. SNR versus BER for DCO-OFDM with FFT size = 64.

References

[1] Light emiting diode www.powerbulbs.com (Accessed: July 1, 2023)

[2] Photo diode parameters www.hamamatsu.com (Accessed: June 28, 2023)

[3] Buchali F, Dischler R and Liu X 2009 Optical OFDM: a promising high-speed optical transport technology *Bell Labs Tech. J.* **14** 125–46

[4] Cailean A-M and Dimian M 2016 Toward environmental-adaptive visible light communications receivers for automotive applications: a review *IEEE Sens. J.* **16** 2803–11

[5] Li C, Yi Y, Lee K and Lee K 2014 Performance analysis of visible light communication using the STBC-OFDM technique for intelligent transportation systems *Int. J. Electron.* **101** 1117–33

[6] Elgala H, Mesleh R and Haas H 2010 An LED model for intensity-modulated optical communication systems *IEEE Photon. Technol. Lett.* **22** 835–7

[7] Lu Q F, Ji X S and Huang K Z 2014 Clipping distortion analysis and optimal power allocation for ACO-OFDM based visible light communication *2014 4th IEEE Int. Conf. on Information Science and Technology (ICIST)* pp 320–3

[8] Farahneh H, Hussain F and Fernando X 2018 Performance analysis of adaptive OFDM modulation scheme in vlc vehicular communication network in realistic noise environment *EURASIP J. Wirel. Commun. Netw.* **2018** 243

[9] Fernando N, Hong Y and Viterbo E 2012 Flip-OFDM for unipolar communication systems *IEEE Tran. Commun.* **60** 3726–33

[10] Goto Y, Takai I, Yamazato T, Okada H, Fujii T, Kawahito S, Arai S, Yendo T and Kamakura K 2016 A new automotive VLC system using optical communication image sensor *IEEE Photon. J.* **8** 1–17

[11] Wu L, Zhang Z, Dang J and Liu H 2015 Adaptive modulation schemes for visible light communications *J. Lightw. Technol.* **33** 117–25

[12] Palomar D P, Cioffi J M and Lagunas M A 2003 Joint Tx-Rx beamforming design for multicarrier MIMO channels: a unified framework for convex optimization *IEEE Trans. Signal Process.* **51** 2381–401

[13] Elgala H, Mesleh R and Haas H 2011 On the performance of different OFDM based optical wireless communication systems *IEEE/OSA J. Opt. Commun. Netw.* **3** 620–8

[14] Turan B, Narmanlioglu O, Ergen S C and Uysal M 2016 Broadcasting brake lights with MIMO-OFDM based vehicular VLC *2016 IEEE Vehicular Networking Conf. (VNC)* pp 1–7

[15] Uysal M, Ghassemlooy Z, Bekkali A, Kadri A and Menouar H 2015 Visible light communication for vehicular need headlamp beam pattern model *IEEE Veh. Technol. Mag.* **10** 45–53

[16] Wang Q, Wang Z and Dai L 2015 Multiuser MIMO-OFDM for visible light communications *IEEE Photon. J.* **7** 1–11

[17] Wang R, Gao Q, You J, Liu E, Wang P, Xu Z and Hua Y 2017 Linear transceiver designs for MIMO indoor visible light communications under lighting constraints *IEEE Trans. Commun.* **65** 2494–508

[18] Ling X, Wang J, Liang X, Ding Z and Zhao C 2016 Offset and power optimization for DCO-OFDM in visible light communication systems *IEEE Trans. Signal Process.* **64** 349–63

[19] Zhang X, Zhou Y F, Yu Y P, Han P C and Wang X R 2014 Comparison and analysis of DCO-OFDM, ACO-OFDM and ADO-OFDM in IM/DD systems *Appl. Mech. Mater.* **701-702** 1059–62

[20] Yu Z, Baxley R J and Zhou G T 2012 EVM and achievable data rate analysis of clipped OFDM signals in visible light communication *EURASIP J. Wirel. Commun. Netw.* **2012** 321

IOP Publishing

Visible Light Communications (Second Edition)
Vehicular applications
Xavier Fernando and Hasan Farahneh

Chapter 6

Precoder and equalizer in 2 × 2 MIMO-VLC systems

Precoding and equalization are important techniques used in MIMO systems to improve performance and mitigate interference. While this area is well researched in RF communications systems, little work has been done on precoding and equalization for optical wireless communication (OWC) systems. That is the focus of this chapter. In a MIMO system, precoding is applied at the transmitter to pre-process the signals before transmission, while equalization is performed at the receiver to compensate for channel effects and recover the original signals. Linear precoding techniques like zero forcing (ZF), minimum mean square error (MMSE) or advanced nonlinear precoding methods like maximum likelihood (MLH) or Tomlinson–Harashima precoding (THP) can be employed to map the data streams from multiple antennas to reduce interference among them. At the receiver, linear equalization techniques such as ZF or MMSE equalization can be applied to the received signals to separate the data streams and combat inter-symbol interference (ISI).

6.1 Precoding basics

6.1.1 Zero forcing

ZF precoding is a spatial signal processing technique used in multi-user multiple-input–multiple-output (MIMO) wireless communication systems to mitigate interference between users. The main goal of this method is to allow a transmitter with multiple antennas to send signals to multiple users simultaneously while minimizing interference between them.

Here's how it works: the transmitter uses knowledge of the channel conditions to each user to design a precoding matrix. This matrix is applied to the signals before transmission in such a way that the signal intended for each user is directed along a path that is orthogonal (perpendicular) to the channels of all other users. As a result,

doi:10.1088/978-0-7503-6049-4ch6

each user receives only their intended signal, with the interference from other users' signals effectively 'forced' to zero.

This technique allows for full spatial multiplexing, meaning multiple data streams can be transmitted simultaneously to different users, increasing overall system capacity. It also takes advantage of multi-user diversity, as the system can choose to serve users with favorable channel conditions. ZF precoding offers a good balance between performance and computational complexity, making it practical for real-world applications. It has been used in long-term evolution (LTE) mobile networks and is being studied for use in massive MIMO systems for future wireless technologies.

However, ZF precoding does have some limitations. It requires accurate channel state information at the transmitter, which can be challenging to obtain in fast-changing environments. Additionally, its performance can degrade in low signal-to-noise ratio (SNR) conditions. Despite these challenges, ZF precoding remains an important technique in modern wireless communications for managing interference and improving spectral efficiency in multi-user MIMO systems.

6.1.2 Minimum mean square error

MMSE precoding is an advanced signal processing technique used in multi-user MIMO wireless communication systems to optimize transmission and mitigate interference between users. Unlike ZF precoding, which completely eliminates interference but may amplify noise, MMSE precoding aims to find an optimal balance between interference suppression and noise enhancement.

MMSE precoding works by designing a precoding matrix at the transmitter that minimizes the mean square error between the transmitted and received signals, taking into account both the channel conditions and the noise at the receivers. This approach allows for better performance in low signal-to-noise ratio (SNR) scenarios compared to ZF precoding.

The MMSE precoder considers the channel state information and the noise power at the receivers to calculate a precoding matrix that optimally distributes the transmit power across the antennas. This results in a trade-off between interference cancellation and noise amplification, leading to improved overall system performance, especially in challenging channel conditions.

The MMSE precoding is particularly useful in scenarios with non-homogeneous user conditions, where users may experience different signal strengths or noise levels. It can adapt to these varying conditions to provide a more balanced and efficient transmission strategy across all users. While MMSE precoding is more computationally complex than simpler techniques like ZF, it offers superior performance in many practical scenarios, making it a popular choice in modern wireless communication systems.

6.1.3 Maximum likelihood

MLH precoding is an advanced signal processing technique used in multi-user MIMO wireless communication systems to optimize transmission and minimize errors. Unlike simpler precoding methods, MLH precoding aims to find the optimal precoding matrix that maximizes the probability of correctly detecting the transmitted signals at the receivers.

The MLH precoder works by considering all possible combinations of transmitted signals and channel conditions, and then selecting the precoding strategy that is most likely to result in correct signal detection at the receivers. This approach takes into account the full statistical properties of the channel and noise, allowing it to achieve optimal performance in terms of minimizing the probability of error. While MLH precoding offers excellent performance, especially in challenging channel conditions, it comes with significant computational complexity. The precoder needs to evaluate a large number of possible signal combinations, which can be computationally intensive, especially as the number of users or antennas increases. Due to this complexity, MLH precoding is often considered impractical for real-time implementations in large-scale systems.

Despite its computational challenges, MLH precoding serves as an important benchmark for evaluating the performance of other precoding techniques. It provides a theoretical upper bound on precoding performance, against which more practical precoding methods can be compared. In scenarios where computational resources are less constrained, or in smaller-scale systems, MLH precoding can offer superior error performance compared to other precoding techniques.

The choice of precoding and equalization techniques depends on factors like channel conditions, computational complexity requirements, and performance targets. Proper design of precoding and equalization allows a MIMO system to leverage spatial multiplexing gains and improve spectral efficiency compared to single-antenna systems. Overall, precoding and equalization work together as key signal processing components to enable reliable high-rate transmission in compact 2 x 2 MIMO deployments.

6.2 Precoding and equalization for VLC

In VLC systems that utilize MIMO technology to enhance data rates and reliability by exploiting spatial diversity, various precoding and equalization techniques are employed to optimize performance.

One prevalent technique is spatial multiplexing, which employs precoding to map multiple data streams onto different transmit LEDs, exploiting the spatial dimension for increased data throughput. This method requires careful design of precoding matrices to mitigate ISI and inter-channel interference, ensuring that each data stream is correctly decoded at the receiver.

Another approach is beamforming, where the transmitter directs the signal towards the receiver using specific beamforming vectors. This technique aims to maximize the received signal power and reduce interference from neighboring LEDs or ambient light sources, thereby enhancing the SNR and improving overall system performance.

Additionally, VLC MIMO systems may utilize equalization techniques such as ZF or MMSE equalization at the receiver. ZF equalization nullifies interference between different streams by inverting the channel matrix, while MMSE equalization minimizes the mean square error between transmitted and received signals, considering noise and channel characteristics.

Furthermore, hybrid techniques like dirty paper coding (DPC) and Tomlinson–Harashima precoding (THP) are gaining traction in VLC MIMO systems. DPC optimally cancels interference from known signals at the transmitter, while THP mitigates noise and improves spectral efficiency by pre-cancelling ISI at the transmitter.

In conclusion, the choice of precoding and equalization techniques in VLC MIMO systems depends on factors such as channel conditions, interference levels, and desired data rates. By leveraging these techniques effectively, VLC MIMO systems can achieve higher spectral efficiency and improved reliability, making them suitable for various indoor communication applications (table 6.1).

Table 6.1. A comparative overview of various precoding and equalization techniques commonly employed in VLC MIMO systems.

Technique	Description	Advantages	Disadvantages
Spatial multiplexing	Precoding maps multiple data streams onto different transmit LEDs, exploiting spatial diversity.	Increased data throughput; utilizes spatial dimension effectively.	Susceptible to ISI; requires accurate channel state information (CSI).
Beamforming	Directs the signal towards the receiver using beamforming vectors.	Enhances signal power; reduces interference from neighboring LEDs.	Requires accurate channel knowledge; limited coverage area due to directional transmission.
Zero forcing (ZF)	Equalization technique that nullifies interference between different data streams.	Removes inter-stream interference; simple to implement.	High computational complexity; amplifies noise and errors in the presence of noise.
Minimum mean square error (MMSE)	Minimizes mean square error between transmitted and received signals considering noise and channel characteristics.	Improved robustness against noise and channel variations; balanced trade-off between performance and complexity.	Higher computational complexity compared to ZF; requires accurate knowledge of channel statistics.
Dirty paper coding (DPC)	Cancels interference from known signals optimally at the transmitter.	Achieves optimal capacity; mitigates interference effectively.	Requires knowledge of interference statistics; complex implementation.
Tomlinson–Harashima precoding (THP)	Pre-cancels ISI at the transmitter, improving spectral efficiency.	Reduces ISI; improves data rates.	Complexity increases with number of transmit antennas; performance depends on accurate CSI.

In this study, we consider three detection schemes, ZF, MLH, and MMSE. Also, we consider flickering/dimming control and the nature of outdoor VLC channel as a frequency selective channel. Moreover, we investigate a transceiver considering perfect Channel State Information (CSI) based on different detection algorithms.

6.2.1 Related work

There is limited published work in precoding and equalization areas in a MIMO-vehicle-to-vehicle (V2V)-VLC System. Most of the existing work on the precoding and equalization of VLC systems is mainly limited to indoor environments. Tomaš *et al* [14] designed a transceiver model for realistic simulation of vehicular VLC. They tested the design of VLC transceiver prototype using a modified version of 802.11 MAC protocol, showing that a MAC protocol specifically designed for VLC should take into account its specific physical properties. In [12], the authors studied a linear transceiver design for indoor VLC with multiple LEDs, they showed that for MIMO-VLC, the optimal precoding reduces to a simple LED selection strategy. Corresponding channel structure and simple receiver design for VLC systems were also discussed in [18].

The MIMO transceiver was optimized to improve the system performance in [7]. The power and the positive offset are jointly designed to improve the spectral efficiency by taking bit error rate (BER) requirement, non-negativity constraint and sum optical power constraint of the transmit signals. The authors in [17] investigated a joint precoding matrix and receiving matrix design via a convergence guaranteed iterative algorithm by taking the positive constraint on the transmit signals into account.

In another work [17], Ying *et al* investigated the design of optical wireless MIMO precoder and equalizer systems for indoor VLC. The authors studied the system with perfect channel state and imperfect channel state information. In [8], a multi-user downlink channel was considered and the corresponding precoding design was optimized. The authors imposed the ZF structure on the precoding matrix. In [7], power and offset allocation with adaptive modulation were well discussed in singular value decomposition (SVD)-based MIMO systems for optical wireless channels, but dimming control was not supported. Authors in [11] proposed a modified SVD-VLC MIMO system to meet the illumination constraints, but additional adjusting should be applied to the input signals so that both the non-negativity of the intensity modulation and dimming control can be satisfied. In the area of a MIMO-VLC system, the precoder and equalizer design are still open to researchers, especially in outdoor VLC applications.

6.2.2 Dimming control technique and its performance in VLC systems

Illumination and communication are the two main functions of the LED lamps in VLC systems. The brightness of the LED light needs to be adjusted in accordance with the requirements and comfort of users and to save energy. The main dimming control technique is pulse width modulation (PWM), where the brightness of the LED light is changed by adjusting the duty cycle of the PWM signal without varying

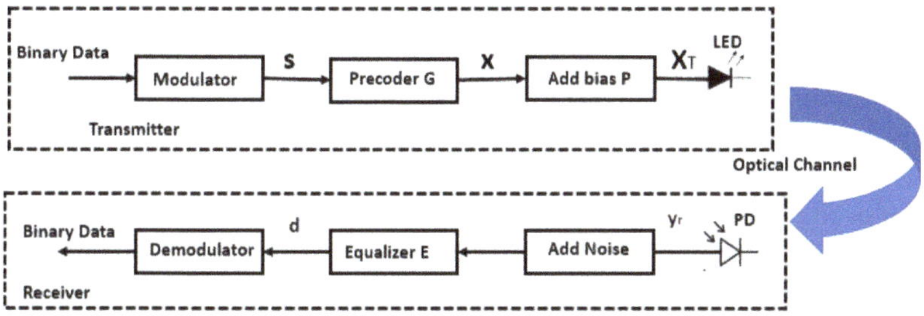

Figure 6.1. Illustration of the investigated precoder and equalizer system.

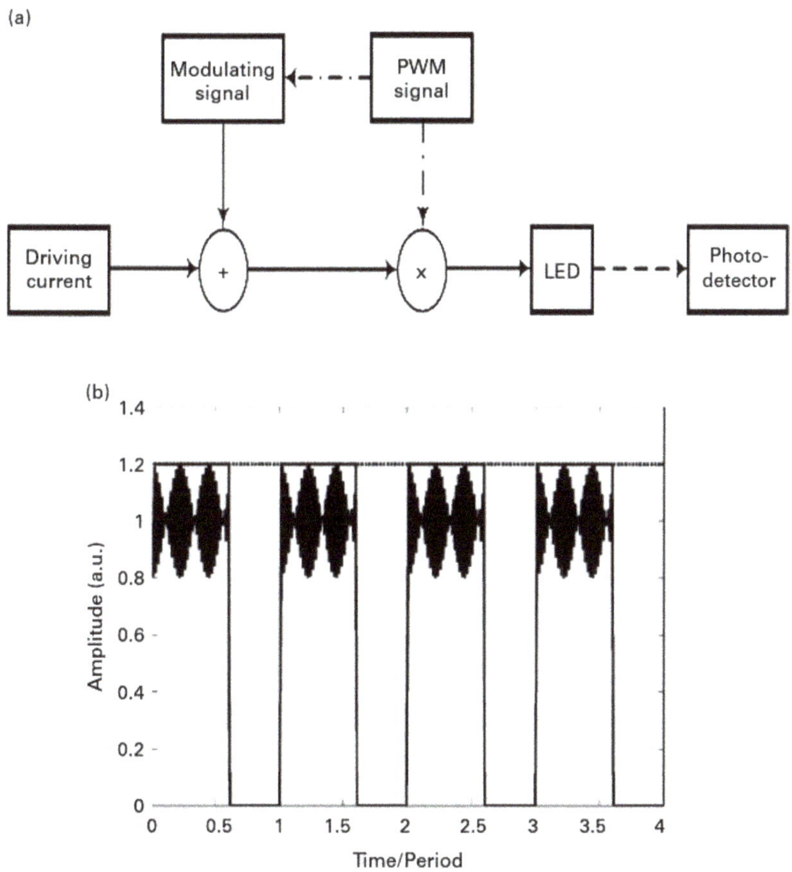

Figure 6.2. (a) Dimming control. (b) Signal waveform with dimming control. Here the duty cycle is 0.6. Reprinted with permission, from [15], copyright (2012) Optical Society of America.

the LED current [15]. As shown in figure 6.2(a), the LED current is modulated by a PWM signal to control its brightness by changing the on duration within the whole period. So the light is dimming during the whole period of the PWM signal. The data are modulated onto the dimming controlled light in the on time only, and no light is transmitted during the off time, as depicted in figure 6.2(b). Since the LED current remains constant throughout, the brightness of the LED light is varied by applying the PWM dimming control signal to adjust the duration of its on period as a fraction of the whole period. With the duty cycle of the PWM signal set to 1, all the LED light is transmitted and the light obtained is of the highest brightness. When the duty cycle is reduced, the LED light in the off period is blocked and hence the light is dimming during the whole duration of the PWM signal. It is worth mentioning that the frequency of the PWM signal should be higher than 200 Hz, otherwise it will cause flicker and would have adverse effects on user health [3, 9].

6.3 Precoder and equalizer in 2 × 2 MIMO V2V-VLC systems

Considering the system given in section 4.3.2 and shown in figure 4.3, the precoder-equalizer system for the V2V-VLC system is shown in figure 6.1, where the binary data modulated, then precoded by the precoder **G**, then DC-biased is added to ensure that the transmitted signal is positive. To this end, the signal is transmitted through an optical channel. In the receiver, the received signal is equalized by equalizer **E**, then passes to be demodulated to predict the transmitted data.

6.3.1 Investigating the transmitter

Since we have 2 × 2 MIMO systems, the original information bits are modulated into source data vector denoted by [5]

$$\mathbf{s}(t) = [s_1(t) \quad s_2(t)]^T \tag{6.1}$$

where any bit s_i is assumed to be return to zero (RZ) with zero mean and two levels $[-A, A]$. The source data vector will be multiplied by a 2 × 2 precoding matrix **G**, where **G** is given as

$$\mathbf{G} = \begin{bmatrix} g_{11} & g_{12} \\ g_{21} & g_{22} \end{bmatrix} \tag{6.2}$$

Each entry of **G** denotes the precoding weight of the transmitted signal between the transmitter s and the receiver r, $(r, s = 1, 2)$.

The precoded data vector **x** is given by

$$\mathbf{x} = \mathbf{G}\,\mathbf{s} \tag{6.3}$$

It is worth noting that since VLC is based on intensity modulation, the vector **x** should be positive. Thus, a DC biasing vector $\mathbf{p} = [p_1 \quad p_2]^T$ is added to the transmitted signal to ensure that the transmitted signal is positive which guarantees that each transmitted vector $(\mathbf{x}_{T_1}, \mathbf{x}_{T_2})$ is positive.

$$\mathbf{x}_T = \mathbf{G}\,\mathbf{s} + \mathbf{p} \tag{6.4}$$

The transmitted data vector $\mathbf{x}_T$ is calculated for the VLC system as follows

$$\begin{bmatrix} x_{T_1} \\ x_{T_2} \end{bmatrix} = \begin{bmatrix} g_{11} & g_{12} \\ g_{21} & g_{22} \end{bmatrix} \begin{bmatrix} s_1 \\ s_2 \end{bmatrix} + \begin{bmatrix} p_1 \\ p_2 \end{bmatrix} \quad (6.5)$$

Let l and u be the lower and the upper levels of the LED dynamic range, respectively. For each LED, we assume that $[l, u](0 < l)$. It is the dynamic-range constraint, and in order to meet the dimming control, i.e., to achieve the illumination requirement by controlling the output power, the following condition must be fulfilled, $l \leqslant x_{T_1},\ \ x_{T_2} \leqslant u$.

Thus, the mean value of any transmitted vector is given by

$$E[x_{T_s}] = \sum_{s=1}^{2}\sum_{r=1}^{2} g_{rs}\ E[s_s] + E[p_s] = p_s \quad (6.6)$$

Equation (6.6) shows that the transmitted signal average is positive and it thus ensures that the transmitted signal is positive, $E[s_s] = 0$ is the mean value of the transmitted bit, $E[p_s]$ is the mean value of the added bias voltage, and s indicates the transmitter and r indicates the receiver with values 1 *or* 2.

To keep the transmitted signal within the LED dynamic range, the following two constraints should be satisfied $\sum_{s=1}^{2}\sum_{r=1}^{2}|g_{rs}|\ A \leqslant u - p$, and $\sum_{s=1}^{2}\sum_{r=1}^{2}|g_{rs}|\ A \leqslant p - l$.

To meet the brightness control of the vehicle's headlights

$$abs(\mathbf{G})\ A \leqslant \min(u - p, p - l) \quad (6.7)$$

The DC bias vector can affect the performance of precoder and equalizer. The BER performance is best when the DC bias is the midpoint of l and u. That is, $p = \frac{1}{2}(l + u)$, we can get a high-performance precoder and equalizer under the constraint of (6.7) [17].

6.3.2 Investigating the receiver

At the receiver, photodiodes (PDs) generate an electric signal proportional to the intensity of the received optical signals. Then, the received signal vector can be expressed as follows [5]

$$\mathbf{y_r} = \gamma\ \mathbf{H}\ \mathbf{x_T} + \mathbf{n} \quad (6.8)$$

where $\mathbf{y}_r$ is the received vectors, $\mathbf{H}$ indicates the channel matrix comprises of line-of-sight (LoS) and non-line-of-sight (NLoS) components and $\mathbf{n}$ is the total noise vector. Here, we consider an additive white Gaussian noise (AWGN) following the distribution of $N(0, \sigma^2 I)$.

For 2 × 2 MIMO, the channel matrix $\mathbf{H}$ is given by

$$\mathbf{H} = \begin{bmatrix} h_{11} & h_{12} \\ h_{21} & h_{22} \end{bmatrix} \quad (6.9)$$

Since we have LoS path and NLoS path coming by the reflectors within the test area, each entry of matrix $\mathbf{H}$ can be given as

$$h(t) = h_{\mathrm{LOS}}(t) + h_{\mathrm{NLOS}}(t) \tag{6.10}$$

Considering a two-dimensional (2D) system, i.e. all sources and receivers lie in the same plane, each entry of the LoS path in **H** is given by

$$h_{rs} = \frac{\cos(\phi_{rs})\cos(\theta_{rs})\, A_r\, T_s(\theta_{rs})\, g(\theta_{rs})}{\pi\, R_{rs}^2}. \tag{6.11}$$

where ϕ_{rs} denotes the emitting angle between source s and receiver r, θ_{rs} denotes the incident angle between source s and receiver r, A_r denotes the effective area of the receiver, $T_s(\theta_{rs})$ denotes the signal transmission coefficient of an optical filter, and $g(\theta_{rs})$ denotes the concentrator gain.

Any entry in the channel matrix **H** for NLoS path between transmitter s and receiver r through reflector q can be calculated as follows

$$h_{rs\ (\mathrm{NLOS})} = h_{qs} + h_{rq} \tag{6.12}$$

It is to be noted that the NLoS gain factor denoted by $h_{rs\ (\mathrm{NLOS})}$ consists of two components, the gain between the transmitter s and reflector q, denoted by h_{qs}, and the gain between the reflector q and the receiver r, denoted by h_{rq}. These gains are calculated as follows

$$h_{qs} = \frac{\cos(\phi_{qs})\cos(\theta_{qs})\, A_q}{\pi\, R_{qs}^2} \tag{6.13}$$

where ϕ_{qs} denotes the emitting angle between source s and reflector surface q, θ_{qs} denotes the incident angle between source s and reflector surface q, R_{qs} denotes the distance between source s and reflector surface q, A_q denotes the effective area of the reflector surface q.

The gain factor h_{rq} is given by

$$h_{rq} = \rho\, \frac{\cos(\phi_{rq})\cos(\theta_{rq})\, A_r\, T_r(\theta_{rq})\, g(\theta_{rq})}{\pi\, R_{rq}^2}. \tag{6.14}$$

In the above equation, ϕ_{rq} is the emitting angle between reflector q and receiver r, θ_{rq} is the incident angle between receiver r and reflector surface q, R_{rq} denotes the distance between receiver r and reflector surface q, and ρ is the reflectivity of the reflector surface. Equation (6.8) can be expanded to two received signals, one from the LoS, and another one from NLoS, $\mathbf{y_r} = y_{\mathbf{r}_{\mathrm{LOS}}} + y_{\mathbf{r}_{\mathrm{NLOS}}}$, where $\mathbf{H} = \mathbf{H}_{\mathrm{LOS}} + \mathbf{H}_{\mathrm{NLOS}}$.

The received signal which is given by (6.8) can be written as

$$\mathbf{y_r} = \gamma\, \mathbf{H}\,(\mathbf{G}\,\mathbf{s} + \mathbf{p}) + \mathbf{n} \tag{6.15}$$

At the receiver side, the term **Hp** is subtracted from $\mathbf{y}_r$ before the equalization. After this subtraction, the communication model can be written as

$$\hat{\mathbf{y}}_{\mathbf{r}} = \gamma\, \mathbf{HG}\,\mathbf{s} + \mathbf{n} \tag{6.16}$$

The estimated symbols by the equalizer can be expressed as follows

$$\begin{aligned} \hat{\mathbf{s}} &= \mathbf{E}(\mathbf{y}_r - \mathbf{HP}) \\ &= \mathbf{E}(\gamma\ \mathbf{HGs} + \mathbf{n}) \end{aligned} \tag{6.17}$$

6.3.3 Equalization matrix design

An equalizer is a digital filter that is used to mitigate the effects of ISI that is introduced by a time-dispersive channel. In order to recover the source data from the received signal, we can formulate the problem as follows 'design a precoder matrix **G** at the transmitter and an equalizer matrix **E** at the receiver to minimize the MSE between the transmitted data and the recovered data', that is,

$$\underset{\mathbf{G},\mathbf{E}}{\text{MSE}} \qquad (\mathbf{d}, \mathbf{s}, \mathbf{G}, \mathbf{E})$$
$$\text{subject to } abs(\mathbf{G})\ A \leqslant \min(u - p, p - l)$$

where $\mathbf{d} = \gamma \mathbf{E}\,\mathbf{H}\,\mathbf{G}\,\mathbf{s} + \mathbf{E}\,\mathbf{n}$.

Here, we assume that the channel matrix **H** is perfectly known. Also, we assume the data vector $\mathbf{s}_s$ is taken from one of an M-pulse amplitude modulation (PAM) symbol with $M = 2^k$ and k is the number of bits per symbol.

6.3.3.1 Zero forcing (ZF) equalizer

The ZF equalizer is a linear equalization algorithm used in communication systems, which inverts the frequency response of the channel. This algorithm applies the inverse of the channel to the received signal, to restore the signal before the channel. In this method, a linear time-invariant filter is used and the ISI component at the output of the equalizer is forced to zero. This will be useful when ISI is significant compared to noise.

The ZF equalization matrix is given by [16]

$$\mathbf{E}_{\mathbf{ZF}} = (\mathbf{H}^{\mathbf{H}}\mathbf{H})^{-1}\,\mathbf{H}^{\mathbf{H}} \tag{6.18}$$

The $\mathbf{E}_{\mathbf{ZF}}$ matrix should have non-zero off diagonal elements and hence this cancels out the interference signal. In fact, it is simple to implement and easy to analyze. The main disadvantage of this equalizer is amplifying the noise. Using ZF equalization, the estimated transmitted symbol vector is [16]

$$\begin{aligned} \hat{\mathbf{s}} &= \gamma\ \mathbf{E}_{\mathbf{ZF}}\ \mathbf{y}_r \\ &= \gamma\ (\mathbf{H}^{\dagger}\,\mathbf{H}\,\mathbf{G}\,\mathbf{s} + \mathbf{H}^{\dagger}\,\mathbf{n}) \\ &= \gamma\ (\mathbf{G}\,\mathbf{s} + \mathbf{H}^{\dagger}\,\mathbf{n}) \end{aligned} \tag{6.19}$$

where $(.)^{\dagger}$ means the pseudo inverse of a matrix.

6.3.3.2 Maximum likelihood equalizer (MLH)

MLH detection calculates the Euclidean distance between the received signal vector and the product of all possible transmitted signal vectors with the given channel **H**, and finds the one with the minimum distance. MLH detection determines the estimate of the transmitted signal vector. The maximum of the likelihood function is achieved when $|\mathbf{y_r} - \mathbf{H}\,\hat{\mathbf{s}}\,|^2$ reaches the minimum value, therefore the estimation of MLH detector is given by [10]

$$\mathbf{s}_{\mathbf{M\hat{L}H}} = \mathbf{arg} \min_{\mathbf{s}_{\mathbf{M\hat{L}H}} \in \mathbf{S}^{\mathbf{M}}} ||\mathbf{y_r} - \mathbf{H}\,\hat{\mathbf{s}}||^2 \tag{6.20}$$

6.3.3.3 Minimum mean squared error equalizer (MMSE)
For MMSE equalizer we have [12]

$$\begin{aligned} & MSE(\mathbf{d}, \mathbf{s}, \mathbf{G}, \mathbf{E}) \\ &= E[||\mathbf{d} - \mathbf{s}||^2] \\ &= E[||(\mathbf{EHG} - I)\,\mathbf{s} + \mathbf{En}||^2] \\ &= Tr((\mathbf{EHG} - \mathbf{I})\mathbf{D}(\mathbf{EHG} - \mathbf{I})^{\mathbf{H}}) + Tr(\mathbf{ER_nE}^H) \\ &= Tr((\mathbf{EHGD})(\mathbf{EHG})^{\mathbf{H}}) + \mathbf{Tr}(\mathbf{ER_nE^H}) + \mathbf{Tr}(\mathbf{D}) - \mathbf{Tr}(\mathbf{EHGD}) - \mathbf{Tr}(\mathbf{D}(\mathbf{EHG})^{\mathbf{H}}) \end{aligned} \tag{6.21}$$

where $\mathbf{D}= E[\mathbf{s}\,\mathbf{s}^H]$, $\mathbf{R}_n = E[\mathbf{n}n^H] = \sigma_n^2\, I_2$, and $\mathbf{D} = \mathrm{diag}\ [D, D]$, where D is given by

$$D = \frac{A^2\,(M + 1)}{3(M - 1)} \tag{6.22}$$

Here, we assume the data vector $\mathbf{s}_s$ is taken from one of a M-PAM symbol with $M = 2^k$ and k is the number of bits per symbol.

The optimal $\mathbf{E}$ for given $\mathbf{G}$ satisfies the following condition

$$\frac{dMSE(\mathbf{d}, \mathbf{s}, \mathbf{G}, \mathbf{E})}{d\mathbf{E}} = 0 \tag{6.23}$$

Then the MMSE equalizer is given by

$$\mathbf{E} = \mathbf{D}\,\mathbf{G}^{\mathbf{H}}\,\mathbf{H}^{\mathbf{H}}\left(\mathbf{H}\,\mathbf{G}\,\mathbf{D}\,\mathbf{G}^{\mathbf{H}}\,\mathbf{H}^{\mathbf{H}} + \sigma_{\mathbf{n}}^2\,\mathbf{I}_2\right)^{-1} \tag{6.24}$$

The related mean squared error (MSE) covariance matrix can be given as

$$\begin{aligned} \mathfrak{R} &= E[(\mathbf{s} - \hat{\mathbf{s}})(\mathbf{s} - \hat{\mathbf{s}})^{\mathbf{T}}] \\ &= (\mathbf{D}^{-1} + \frac{1}{\sigma_{\mathbf{n}}^2}\mathbf{G}^{\mathbf{H}}\,\mathbf{H}^{\mathbf{H}}\,\mathbf{H}\,\mathbf{G})^{-1} \end{aligned} \tag{6.25}$$

6.3.4 Precoding matrix

6.3.4.1 MMSE precoder
Since MMSE equalization technique is investigated, here we will use the corresponding precoder matrix which is given by [13]

$$\mathbf{G}_{\mathrm{MMSE}} = \sqrt{\frac{1}{2}}\,\mathbf{V}\ \triangle\ \tilde{\mathbf{D}} \tag{6.26}$$

where $\mathbf{V}$ comes from the decomposition of channel matrix $\mathbf{H}^*\Phi_{nn}^{-1}\mathbf{H}$ and $\mathbf{H} = \mathbf{V}\,\boldsymbol{\Theta}\,\mathbf{V}^*$, $\boldsymbol{\Phi}_{nn} = 2\,\sigma_n^2\, I$ is the noise matrix, $\tilde{\mathbf{D}}$ is a normalized discrete Fourier transform matrix, and $\boldsymbol{\Delta}$ is a diagonal matrix with the diagonal elements determined by [4]

$$\|m_{ii}\| = \frac{1}{2}\left(1 + \sum_{j=1}^{2}\theta_j^{-1}\right) - \theta_i^{-1} \tag{6.27}$$

where θ_i are diagonal elements of Θ coming from the eigenvalue decomposition $\mathbf{H}^*\,\Phi_{nn}^{-1}\,H = \mathbf{V}\,\boldsymbol{\Theta}\,\mathbf{V}^*$.

6.3.4.2 ZF precoder

The precoder for ZF equalization is [16]

$$\mathbf{G}_{\mathbf{ZF}} = \sqrt{\frac{1}{2\,\mathbf{tr}(\Theta^{-1/2})}}\,\mathbf{V}\,\Theta^{-1/4}\,\tilde{\mathbf{D}} \tag{6.28}$$

6.3.4.3 Maximum likelihood (MLH) precoder

The precoder for MLH equalization is given by [10]

$$\mathbf{G}_{\mathbf{MLH}} = \begin{bmatrix} g_{m11} & g_{m12} \\ g_{m21} & g_{m22} \end{bmatrix} \tag{6.29}$$

subject to the following constraints $g_{m11}\,g_{m21} + g_{m12}\,g_{m22} = 0$ and $g_{m11}^2 + g_{m12}^2 + g_{m21}^2 + g_{m22}^2 = 0$.

6.4 Simulation and results

In this study, we use MATLAB to model the system and simulation parameters are summarized in tables 2.4 and 4.1. In the simulations, we consider multipath propagation with one reflection system where all reflectors surfaces are assumed to have same reflective factor. Moreover, without loss of generality, we consider a clear sky and the vehicles are moving at uniform speed in the same direction. We use a Philips Ultinon LED 12985BWX2 as transmitter [1]. Also, we use an avalanche photodiode (APD) model (Si) Hamamatsu S8664-1010 as a receiver [2]. Moreover, we also consider 4-PAM modulated symbols and the constellation of 4-PAM is formed in the range of $[-3,\ 3]$ (i.e. $A = 3$). The investigated optimal precoder and equalizer can be easily applied to the practical design. We generate and fix the NLoS gain as given by (4.40). Both the signals and the noises are assumed to be independent, which is also reasonable in V2V-VLC practical systems. Additionally, we consider the ambient noise model used in [6]. To measure the performance of the different schemes, we compare the performance of the investigated precoders and equalizers in terms of BER and MSE. Note that the power constraint in (6.7) is still valid for all the obtained simulations [5].

Figure 6.3 shows the comparison of the investigated schemes from BER standpoint. The MMSE precoding scheme shows better improvement in terms of BER.

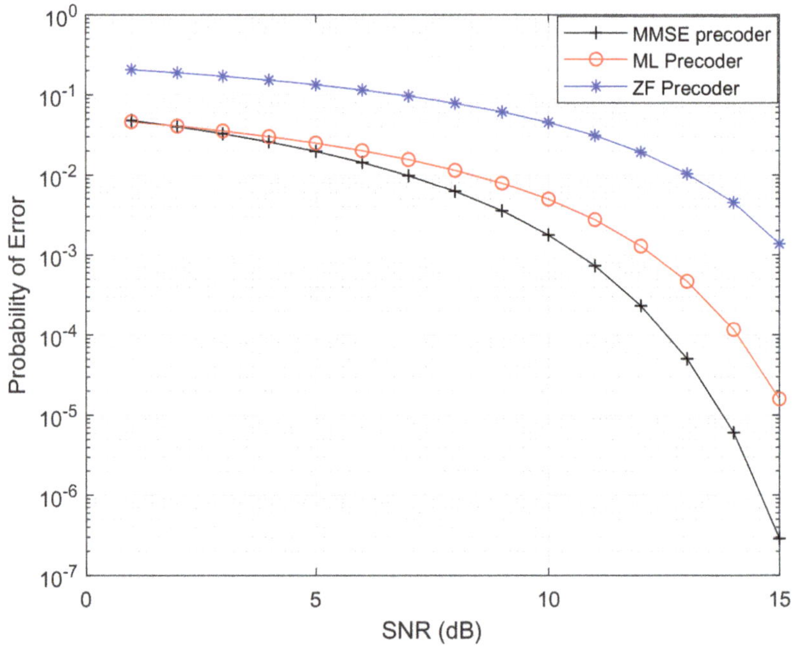

Figure 6.3. Comparison of the suggested detection algorithms in terms of BER. Copyright 2018, IEEE. Reprinted, with pemission, from [5].

For MMSE and MLH detectors, we can see here that performance curves of these two systems are close to each other when SNR is low, but the gap gets larger when SNR gets higher, while the performance of ZF shows a gap with MLH and MMSE for all values of SNR. For high SNR the gap is about 2 dB between MMSE and MLH, while the gap between ZF and the two other schemes is about 8 dB. When SNR is less, that means noise is large and the different detection schemes have almost the same behavior. When the SNR is large, the post detection of SNR may have been affected by channel matrix **H**.

Figure 6.4 shows a comparison of the suggested detection algorithms in terms of MSE. The MSE is a measure of the quality of an estimator, it is always non-negative as shown in figure 6.4. However, the MSE values for all detection schemes are very close and very low, which means those schemes can be valid to use with a V2V-VLC system as the values closer to zero are better.

In general, MMSE shows better performance than MLH and ZF because the MMSE detector is dependent on the noise variance and the condition number of the channel matrix,

6.5 Chapter summary

This chapter investigates the precoding and equalizing scheme for the V2V-VLC system considering flickering/dimming control and the nature of the outdoor VLC channel as a frequency selective channel to enhance the system's performance. We present the precoder matrix, as well as the equalizer matrix. We compare the system

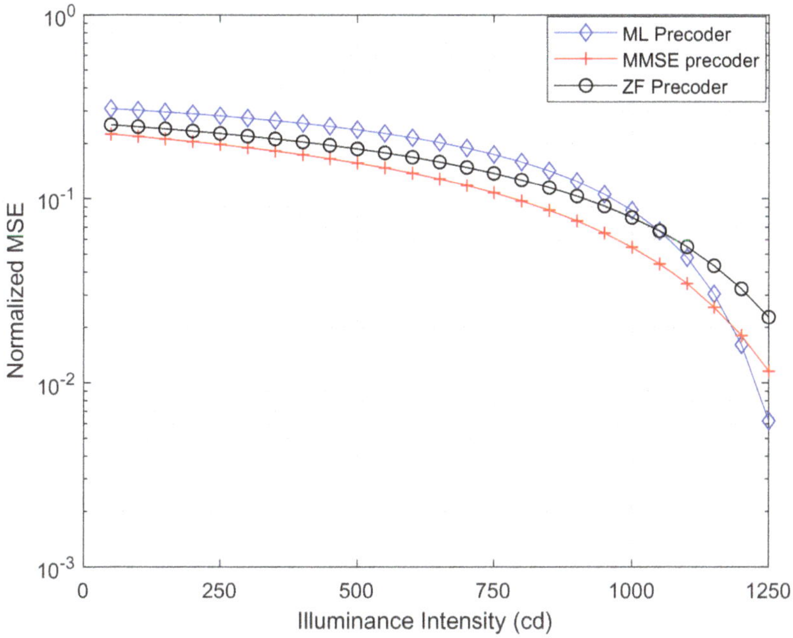

Figure 6.4. Comparison of the suggested detection algorithms in terms of MSE. Copyright 2018, IEEE. Reprinted, with pemission, from[5].

performance of MMSE, ZF, and MLH precoders, where the MMSE scheme shows a noticeable improvement.

References

[1] Light emiting diode. available online: www.powerbulbs.com (Accessed: July 1, 2023)

[2] Photo diode parameters. available online: www.hamamatsu.com (Accessed: June 28, 2023)

[3] Arnon S 2015 *Visible Light Communication* (Cambridge: Cambridge University Press)

[4] Fang D 2013 Optimal precoder design and block-equal QRS decomposition for ML based successive cancellation detection *Master Thesis* McMaster University

[5] Farahneh H, Hussain F, Hussain R and Fernando X 2018 A novel optimal precoder and equalizer in 2 × 2 MIMO VLC systems for vehicular application *2018 IEEE Globecom Workshops (GC Wkshps)* pp 1–6

[6] Fernando X, Farahneh H and Kamruzzaman S 2018 Differential receiver as a denoising scheme to improve the performance of V2V-VLC systems *ICC 2018-USA* vol 10 pp 45–53

[7] Park K-H, Ko Y-C and Alouini M S 2013 On the power and offset allocation for rate adaptation of spatial multiplexing in optical wireless MIMO channels *IEEE Trans. Commun.* **61** 1535–43

[8] Ma H, Lampe L and Hranilovic S 2013 Robust MMSE linear precoding for visible light communication broadcasting systems *2013 IEEE Globecom Workshops (GC Wkshps)* pp 1081–6

[9] Ntogari T, Kamalakis G, Walewski T and Sphicopoulos J 2011 Combining illumination dimming based on pulse-width modulation with visible-light communications based on discrete multitone *IEEE/OSA J. Opt. Commun. Netw.* **3** 56–65

[10] Ozaki T, Iwai H and Sasaoka H 2014 Improvement of transmission performance by variable precoding in MIMO transmission *IEICE Commun. Express* **3** 168–74
[11] Elgala H, Butala P M and Little T D C 2013 SVD-VLC: a novel capacity maximizing VLC MIMO system architecture under illumination constraints *Proc. IEEE Globecom Workshop OWC* vol 1 pp 1087–92
[12] Wang R, Gao Q, You J, Liu E, Wang P, Xu Z and Hua Y 2017 Linear transceiver designs for MIMO indoor visible light communications under lighting constraints *IEEE Trans. Commun.* **65** 2494–508
[13] Davidson T N, Chan S S and Max Wong K 2004 Asymptotically minimum BER linear block precoders for MMSE equalisation *Communications, IEE Proc.* vol 151 pp 297–304
[14] Tomaš B, Tsai H M and Boban M 2014 Simulating vehicular visible light communication: physical radio and MAC modeling *2014 IEEE Vehicular Networking Conf. (VNC)* pp 222–5
[15] Wang Z, Zhong W-D, Yu C, Chen J, Francois C P S and Chen W 2012 Performance of dimming control scheme in visible light communication system *Opt. Express* **20** 18861–8
[16] Ding Y, Davidson T N, Luo Z-Q and Wong K M 2003 Minimum BER block precoders for zero-forcing equalization *IEEE Trans. Signal Process.* **51** 2410–23
[17] Ying K, Qian H, Baxley R J and Yao S 2015 Joint optimization of precoder and equalizer in MIMO VLC systems *IEEE J. Sel. Areas Commun.* **33** 1949–58
[18] Zeng L, O'Brien D C, Minh H L, Faulkner G E, Lee K, Jung D, Oh Y and Won E T 2009 High data rate multiple input multiple output (MIMO) optical wireless communications using white LED lighting *Sel. Areas Commun.* **27** 1654–62

IOP Publishing

Visible Light Communications (Second Edition)
Vehicular applications
Xavier Fernando and Hasan Farahneh

Chapter 7

Sunlight effects and denoising schemes

In this chapter, the effect of solar irradiance and the impact of noise from other external sources are investigated for a vehicle-to-vehicle (V2V) visible light communication (VLC) system with regard to signal-to-noise ratio (SNR), bit error rate (BER), and data rate. Then, we present two schemes to combat the effect of the ambient light on the V2V-VLC system. Firstly, we present the differential receiver as an efficient denoising scheme that cancels out the common mode (ambient) noise. This differential receiver exploits multiple wavelengths. Then, we propose a machine learning-based adaptive k-Nearest Neighbor (kNN) scheme to dynamically minimize the sunlight noise and maximize SNR by controlling both the transmitter irradiance and the receiver field of view (FoV). The second approach, combined with the extended Kalman filter approach proposed in Chapter 3 is very promising.

7.1 Solar irradiance

It is a well-known fact that 47% of the total solar irradiance falls within the visible light frequency band of the spectrum [16], as shown in figure 7.1. The main meteorological conditions in network locations determine the level of sunlight. Therefore, the major challenge of the outdoor-VLC system is the strong influence of the ambient-light noise because of sunlight. Sunlight represents unmodulated sources, that can be received at an average power much larger than the desired signal, even when optical filtering is employed. However, the noise due to the solar irradiance and other surrounding light sources, is a major concern that degrades the performance of the VLC system in outdoor application in terms of SNR and BER. We have studied minimization of interference using adaptive tracking with the extended Kalman filter approach in chapter 2. We will look into a different approach in this chapter.

Solar irradiance is defined as the amount of solar power in any place. The Earth's rate of rotation, the difference in local longitude, and the standard meridian for the local time zone are considered the main keys in irradiance calculations. Measuring

doi:10.1088/978-0-7503-6049-4ch7

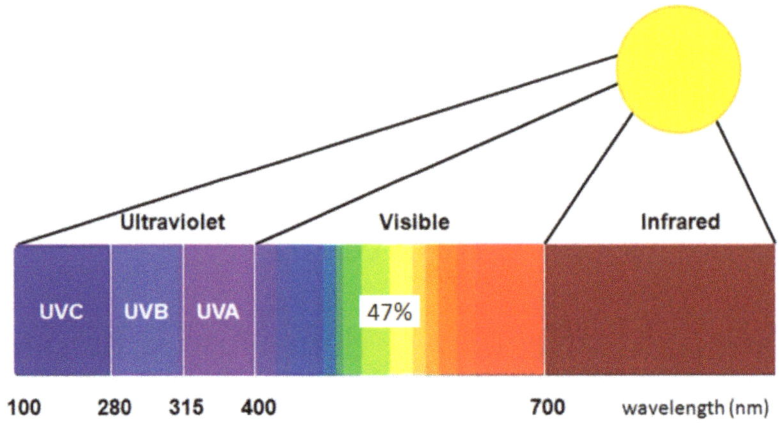

Figure 7.1. Sunlight spectrum.

irradiance correctly needs highly accurate apparatus. In general, solar irradiance can be calculated from clear sky solar radiation on a horizontal surface in W m^{-2} [6].

The most important parameter to measure direct solar radiation (irradiance) is q factor and is given by [6]

$$q_{\text{sun}} = \left(1350.3\left[1 + 0.099\cos\left(\frac{360n}{365}\right)\right][\sin\varphi\sin\zeta + \cos\varphi\cos\zeta\cos\omega]\right) \quad (7.1)$$

where ζ is solar declination, ω is the angular displacement, φ is the location longitude, and n is the day of the year.

However, calculating solar irradiance is beyond the scope of this book, instead, the effect of solar irradiance on the V2V-VLC system will be addressed.

7.2 SNR of V2V-VLC system

In this study, we consider the system given in section 4.3.2. Also, we chose the city of Toronto as a reference to calculate solar irradiance during a full year (12 months) period to cover the worst and best cases of irradiance. We assume the sunlight incident angle within the FoV of the receiver all the time. Furthermore, an on–off keying (OOK) modulation scheme is used. Table 7.1 gives classes of natural light in different metrological conditions and at different times while the values of solar irradiance are given in table 7.2.

Figure 7.2 shows the scenario of solar irradiance effect on V2V-VLC systems. It can be seen that interference due to solar irradiance is the major obstacle on a bright day.

Figure 7.3 shows the response of the avalanche photodiode (APD) of predicted spectral irradiance of the sunlight for different visible light wavelength in the city of Toronto. Here, APD response is given without any optical filter scheme. To show the strong effect of the solar irradiance on the V2V-VLC system, we calculate the SNR for the V2V-VLC system without any denoising scheme. Figure 7.4 shows the strong effect of sunlight without any filtering scheme on the SNR of the system

Table 7.1. Classes of natural light. Reprinted with permission from [1].

Natural light class	Intensity (lx)
Sunlight not considered	0
Clear night, full moon	0.3
Winter's day, overcast sky	900–2000
Summer's day, overcast sky	4000–20 000
Winter's day, clear sky	Up to 9000
Summer's day, clear sky	Up to 100 000

Table 7.2. Solar irradiance at the city of Toronto. Reprinted with permission from [1].

Month	Average Solar Insolation Measured in kWh m^{-2} day^{-1} with solar panel direction southwest 45° from south
January	1.5
February	2.35
March	3.29
April	4.35
May	5.12
June	5.88
July	5.87
August	5.02
September	3.92
October	2.64
November	1.55
December	1.24

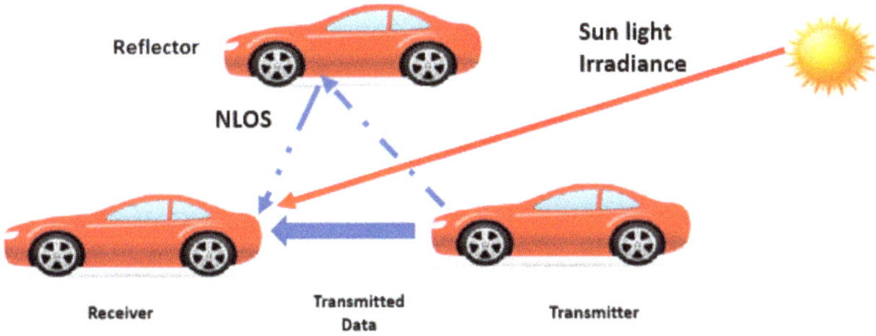

Figure 7.2. The effect of solar irradiance on V2V-VLC systems.

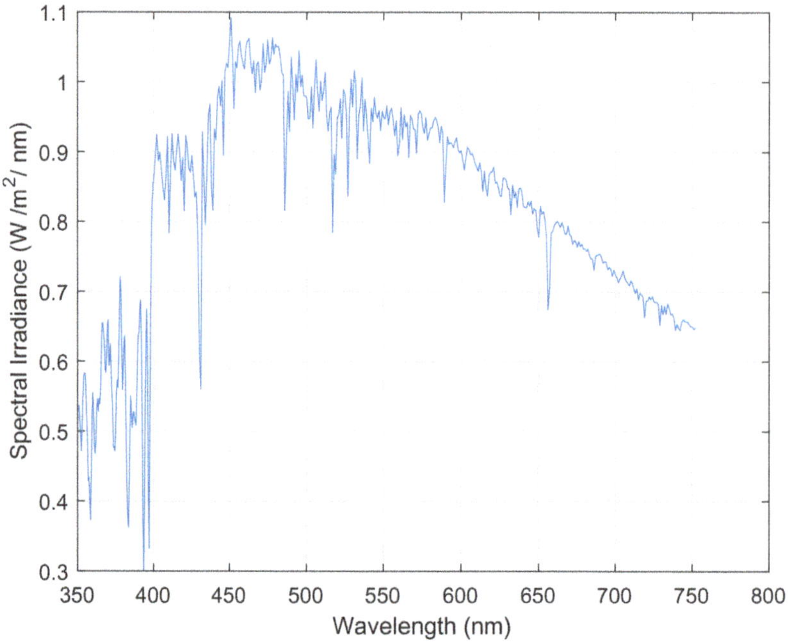

Figure 7.3. Average predicted solar irradiance at the city of Toronto in month of June [3, 7]. Copyright 2018, IEEE. Reprinted, with permission, from [7].

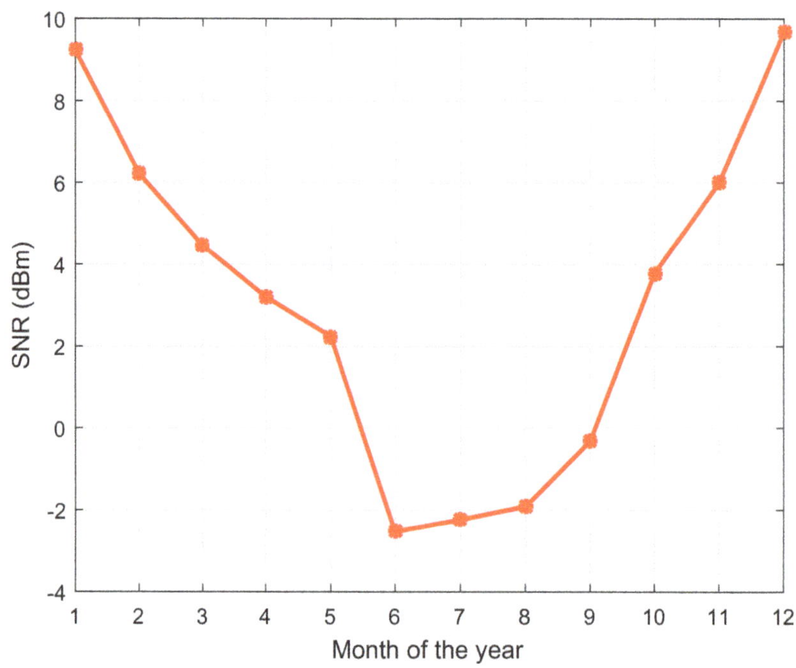

Figure 7.4. SNR without any filtration scheme at the city of Toronto when the LED illuminance intensity is 1750 (cd) [3, 7]. Copyright 2018, IEEE. Reprinted, with permission, from [7].

during different months of the year. The summer's months show the worst SNR, where SNR approaches ≃3 dBm, and in winter, SNR approaches ≃9 dBm.

7.3 Related work in denoising schemes for VLC

There are different approaches for denoising in optical systems. These are shown in table 7.3. However, many approaches deal with the received signal at the sample level. These are not practically applicable for high-speed vehicular communications. Hence, we focus only on techniques that are feasible for high-speed vehicle-to-everything (V2X) environments.

Most of the previous works for the outdoor-VLC system did not consider the effect of a practical ambient-light noise which varies from time to time throughout the year, and from location to location. Few studies have addressed the effect of sunlight on an outdoor-VLC system. In [16], the authors considered the impact of sunlight as a function of location, time and for different surfaces over the four seasons of the year and analyzed the solar irradiance. Moreover, they investigated the effect of solar irradiance in the outdoor-VLC system in terms of data rate and BER degradation. The authors in [14] presented a daylight noise model based on a modified blackbody radiation model to capture the effect of ambient-light noise. Also, they presented a new receiver using the selective combining technique to reduce the effect of background noise. The authors in [15] proposed a measurement-based time variant non-clear sky channel model for an internet-to-vehicle (I2V)-

Table 7.3. Brief comparison of denoising ambient-light noise in VLC systems.

Scheme	Advantages	Disadvantages
Linear filtering	Simple, real-time processing, fixed parameters	Possible signal distortion
Nonlinear filtering	Effective for impulsive noise, adaptive	More complex, may not preserve edges well
Wavelet transform	Multi-resolution analysis, effective noise removal	Computationally intensive, may not be suitable for real-time applications, sensitive to parameters
Deep learning-based	Adaptive learning, high performance	Requires large training data, high computational resources, may not be suitable for real-time applications
Kalman filtering	Optimal estimation, adaptive	computational complexity
Compressed sensing	Efficient sampling, effective denoising	Complex reconstruction, sensitivity to sparsity
Proposed differential filtering	Effective for common mode noise in a MIMO system	Multiple wavelengths are involved

VLC system which considers the dynamic characteristics of background radiation to enable more realistic and accurate prediction of the VLC system performance on the outdoor application. Moreover, they introduced a new receiver design with dual-reception and effective ambient-light rejection capabilities which employs the selection diversity technique in order to mitigate the impact of ambient-light noise due to daylight. In [8], the authors presented an analytical analysis of solar irradiance. They investigated the effect of solar irradiance on the VLC system by using an optical filter in terms of data rate and BER degradation.

The authors in [12] showed the applicability and feasibility of different machine learning (ML) techniques based on iris recognition through smartphone-captured images. Authors trained different classifiers and used histogram equalization processes to maximize accuracy. In [17], the authors proposed a novel object tracking framework based on an online learning scheme which can work robustly in challenging scenarios. They also proposed, a learning-based particle filter with color and edge-based features. Moreover, a multiple classifiers fusion localization technique using the received signal strengths of visible light is proposed in [18], in which LEDs are used to transmit various intensity modulated sinusoidal signals to be received by PDs placed at various grid points.

7.4 Noise calculations

At the receiver, the received signal consists of the transmitted signal and the noise signal. The photocurrent I_o due to the received optical power consists of two components, the line-of-sight (LoS) component and the non-line-of-sight (NLoS) component, i.e, ($I_o = I_{o_{\mathrm{LOS}}} + I_{o_{\mathrm{NLOS}}}$) and is given as

$$I_o = G\,A_r\left(\int_{\lambda_1}^{\lambda_2} P_r^{\mathrm{LOS}}(\lambda)\,\gamma(\lambda)\,T_o(\lambda)\,d\lambda + \int_{\lambda_1}^{\lambda_2} P_r^{\mathrm{NLOS}}(\lambda)\,\gamma(\lambda)\,T_o(\lambda)\,d\lambda\right) \tag{7.2}$$

where $\lambda_1,\ \lambda_2 \in [350\ \ 750]$ (nm) are the visible light wavelengths, $P_r^{\mathrm{LOS}}(\lambda),\ \ P_r^{\mathrm{NLOS}}(\lambda)$ are the received power and are given in (4.42), and (4.43), respectively, for certain wavelength λ, G is APD gain, $\gamma(\lambda)$ is the responsivity of the APD given in A/W, A_r is the effective area of the receiver, and $T_o(\lambda)$ is the transmittance of the bandpass optical filter.

Similarly, the photocurrent I_s produced at the APD due to the solar irradiance is given as,

$$I_s = G\,A_r\,\cos(\theta_{\mathrm{sun}})\int_{\lambda_1}^{\lambda_2} P_s(\lambda)\,\gamma(\lambda)\,T_o(\lambda)\,d\lambda \tag{7.3}$$

where $P_s(\lambda)$ is the solar irradiance given in W m^{-2} nm^{-2}, θ_{sun} is the incident angle of the sunlight on the receiver surface.

The random arrival of the incident photons from both LED light, and sunlight result in shot noise. This type of noise can be modeled by Poisson process. Moreover, when the number of incident photons is large, shot noise is approximated by a Gaussian process (using the central limiting theorem).

According to [13] the shot noise variance is given as

$$\sigma^2{}_{\text{Shot}} = 2\, q_e\, G^2\, F\, (I_o + I_s)\, B \tag{7.4}$$

where q_e is the electron charge, F is the excess noise; $F = k_n\, G + (2 - \frac{1}{G})(1 - k_n)$ where k_n is the holes/electrons ionization rate [10], and B is the APD bandwidth.

7.4.1 Other noise sources

In addition to the shot noise, there are many different types of noises degrading the performance of the VLC system such as dark noise. Dark noise is caused due to avalanche effect of the APD and is given as [13]

$$\sigma^2{}_{\text{Dark}} = 2\, q_e\, G^2\, F\, I_{dg}\, B + 2\, q_e\, I_{ds} \tag{7.5}$$

where I_{ds} is the surface dark current and I_{dg} is the bulk dark current that experiences the avalanche effect of the APD.

Also, due to the random thermal motion of charge carriers, thermal noise is generated, the variance of thermal noise is given as [13]

$$\sigma^2{}_{\text{Thermal}} = 4 \left(\frac{K_{\text{B}}\, T_k}{R_L} \right) F_n\, B \tag{7.6}$$

where K_{B} is Boltzmann constant, T_k is the temperature in Kelvin, R_L is the load resistance given as 50 Ω, and F_n is the photodiode noise figure.

Due to variances in the summation principle which is given in section 1.3, the total noise variance is given by

$$\sigma^2{}_n = \sigma^2{}_{\text{Shot}} + \sigma^2{}_{\text{Dark}} + \sigma^2{}_{\text{Thermal}} \tag{7.7}$$

In this study, we consider Hamamatsu S8664 model for APD [3]. The parameters and their values are given in table 2.4.

7.4.2 Performance metrics

The SNR for the system is,

$$SNR = \frac{I_o}{\sigma^2{}_n} \tag{7.8}$$

The maximum achievable data rate defined as,

$$R_{\text{max}} = B \log_2 (1 + SNR) \tag{7.9}$$

For OOK modulation scheme, BER for certain SNR is given by equation (4.48).

7.5 Differential receiver as a denoising scheme to improve the performance of V2V-VLC Systems

Achieving high SNR, high data rate, and minimizing multipath dispersion are the most challenges during the receiver design for any system. In order to design an efficient optical filtering system for the outdoor V2V-VLC system in the presence of

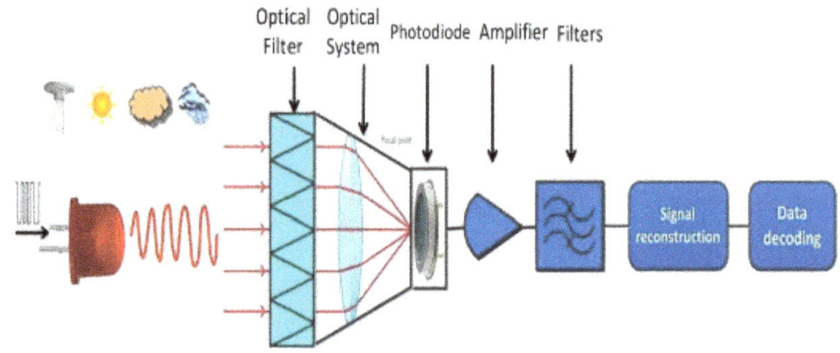

Figure 7.5. The receiver architecture of the VLC system.

sunlight and other artificial lights, eye safety should additionally be taken into account. Many previous works considered bandpass blue optical filter for denoising process [4, 8]. Such schemes include APD with collimate lens on the top of the APD to focus incident light onto the surface of the APD. The optical filter is laid on this lens for blocking unwanted lights, as shown in figure 7.5.

In this study, a differential receiver with optical filter scheme is presented as a solution to combat the effect of the sunlight noise. We investigate the system performance for three scenarios: without optical filter, with optical filter, and with the proposed differential receiver.

7.5.1 Differential filtering scheme for 2 × 2 MIMO-V2V-VLC

Differential filtering scheme is a well-known idea in communication field [9, 11]. In this study, we apply this scheme with some modifications to the system to mitigate the effect of sunlight noise. The differential filtering circuit is shown in figure 7.6. Each transmitter source transmits with different wavelength, i.e., transmitter 1 uses λ_1 and transmitter 2 uses λ_2, where the difference between λ_1 and λ_2 is several tens of nanometers, thus, we can assume the channel coefficients for two transmitters are equal. We consider red shift phenomena to adjust the wavelengths λ_1 and λ_2 [5].

For 2 × 2 MIMO, the received signal at any receiver $\mathbf{y}_{ri}$ is give by

$$\mathbf{y}_{ri} = [h_{i1} \quad h_{i2}][\mathbf{x}_1 \quad \mathbf{x}_2]^T + \mathbf{n}_i \tag{7.10}$$

where $(i = 1, 2)$, $\mathbf{x}_1$, $\mathbf{x}_2$ indicate the transmitted vector of first and second transmitter, respectively, with wavelength λ_1, and λ_2. $\mathbf{n}_i$ indicates the noise vectors including shot noise, which is rising up from the sunlight in addition to thermal noise and dark noise. We will assume only the shot noise, as it is the dominant noise.

The received signal is collected by a lens as shown in figure 7.6. Thereafter, the signal passes to a spatial filter to alter the structure of light. Then, the signal passes through two parallel bandpass optical filters with equal bandwidth but different center frequencies ($f_1 = \frac{c}{\lambda_1}$, $f_2 = \frac{c}{\lambda_2}$), where c is the speed of light. Afterward, the outputs of the optical filters pass to APDs and then the signal passes through a low pass filter (LPF). At the end, two parallel signals $\mathbf{y}_1$, and $\mathbf{y}_2$ are subtracted by an

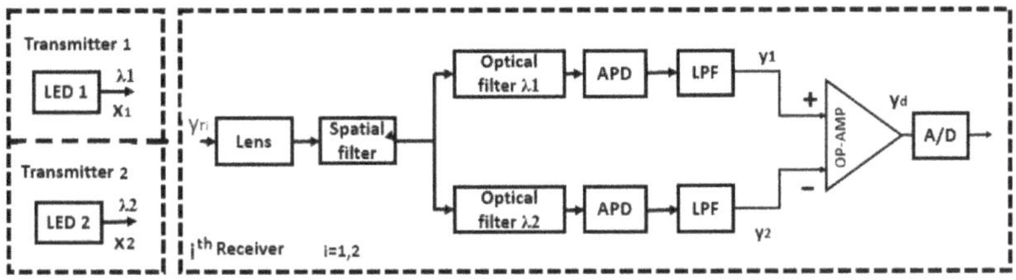

Figure 7.6. Denoising differential scheme of VLC.

operational amplifier (OP-AMP) to form one signal ($\mathbf{y}_d = \mathbf{y}_1 - \mathbf{y}_2$) to pass to an analog-to-digital (A/D) converter, and finally to the signal detection process [7].

Since the optical filter will pass either $\mathbf{x}_1$, or $\mathbf{x}_2$ depending on the signal wavelength, $\mathbf{y}_1$, and $\mathbf{y}_2$ are given by

$$\mathbf{y}_1 = \mathbf{H}_T\,\mathbf{x}_1 + \mathbf{n}_1 \tag{7.11}$$

$$\mathbf{y}_2 = \mathbf{H}_T\,\mathbf{x}_2 + \mathbf{n}_2 \tag{7.12}$$

where $\mathbf{H_T}$ is the transfer function of the channel of one branch of the differential receiver. In this study, $\mathbf{H_T}$ can be considered the same for two parallel branches.

The output of the OP-AMP, $\mathbf{y_d}$ is given by

$$\begin{aligned}\mathbf{y_d} &= \mathbf{y}_1 - \mathbf{y}_2\\ &= \mathbf{H}_T\,\mathbf{x_d} + \mathbf{n_d}\end{aligned} \tag{7.13}$$

where ($\mathbf{x}_d = \mathbf{x}_1 - x_2$), and ($\mathbf{n}_d = \mathbf{n}_1 - \mathbf{n}_2$).

7.5.2 Signal filtration by differential receiver

Assuming the sunlight noise is a stochastic stationary random process. The background noise component after the LPF will be cancelled due to the differential detection system.

The sunlight noise spectrum, whose envelope denotes by S, is spread over a very large spectral band B_s. Because we assume each headlight transmits the signal with different λ, we need two bandpass filters (BPFs) (optical filters) at the suggested differential detector with central frequencies f_1 and f_2. Their impulse response functions are $h_1(t)$ and $h_2(t)$, respectively, and the corresponding transfer functions are $H_1(f)$ and $H_2(f)$, respectively. Also, let the two filters have the same bandwidth given by B_f. Since $B_s > > B_f$, the output of the filters S_1 and S_2 can be considered flat.

Assuming the optical filter is a linear time-invariant system, S_1 *and* S_2 are given by [7]

$$S_i(t) = \int_{-\infty}^{\infty} h_i(\tau)\, S(t - \tau)\, d\tau \qquad i = 1, 2. \tag{7.14}$$

The filters output S_1 and S_2 are passed to the APD. At the output of ith APD, the light intensity is given by

$$I_{Si}(t) = \int_{T_s} |S_i(t)|^2 \, dt \tag{7.15}$$

where T_s is the symbol period of the assigned modulation scheme.

Hence, $|S_1(t)|^2$ is given by

$$|S_1(t)|^2 = \int_{-\infty}^{\infty} \int_{-\infty}^{\infty} h_1(\tau_1)\, S(t - \tau_1)\, h_1(\tau_2)\, S(t - \tau_2) d\tau_2 \, d\tau_1 \tag{7.16}$$

For first APD, substitute (7.16) in (7.15), we get

$$I_{S1}(t) = \int_{-\infty}^{\infty} \int_{-\infty}^{\infty} h_1(\tau_1)\, h_1(\tau_2) \int_{T_s} S(t - \tau_1)\, S(t - \tau_2) d\tau_2 \, d\tau_1 \, dt \tag{7.17}$$

Since S is considered to be an ergodic stationary random process, then the third integral in (7.17) is the autocorrelation function of $S(t)$. Thus, we can write it in terms of expected value $E[.]$ as

$$E(S(t - \tau_1)\, S(t - \tau_2)) = \int_{T_s} S(t - \tau_1)\, S(t - \tau_2)\, dt = R_s(\tau_2 - \tau_1) \tag{7.18}$$

Assuming that H_1 is an ideal BPF, its impulse response is $h_1(t) = \text{sinc}(2\pi B_f t)\cos(2\pi f_1 t)$. The impulse response $h_1(t)$ is a sinc function with its first zero at $\frac{1}{\Delta f}$ ($\Delta f = f_2 - f_1$), modulated by $\cos(2\pi f_1 t)$. On the other hand, assuming a rectangular-shaped power spectral density (PSD) for S (just for the sake of demonstration simplicity), its autocorrelation function will be a sinc function with its first zero at $\frac{1}{B_S}$. Given that $B_s >> \Delta f$ in practice, this autocorrelation can relatively be considered as a Dirac delta function. In other words, S can be considered as a white noise in the calculation of I_{S1}. Figure 7.7 shows an illustration for the PSD of the background noise. Also, it shows bandpass filtering of bandwidth

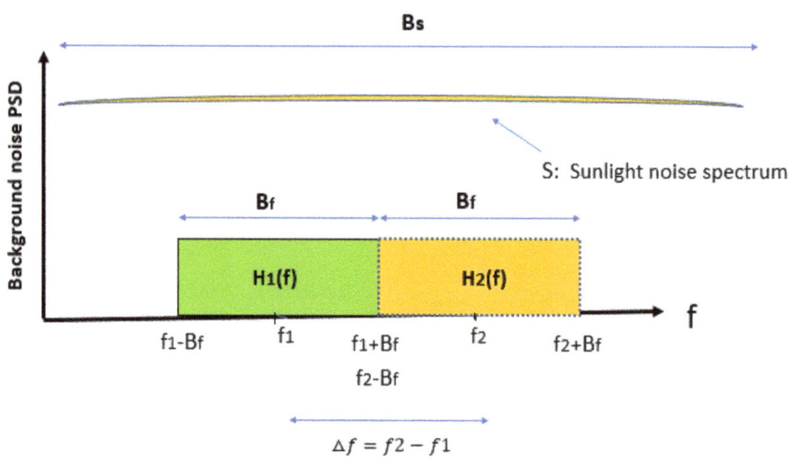

Figure 7.7. Background noise PSD.

B_f around f_1 and f_2, corresponding to the bandpass background noises' field envelopes S_1 and S_2.

The autocorrelation function in (7.18) can be given by

$$R_s(\tau_2 - \tau_1) = \sigma_s^2\, \delta(\tau_2 - \tau_1) \tag{7.19}$$

where σ_s^2 is the variance of the noise and δ is Dirac delta function.

Furthermore, (7.15) can be written as

$$I_{S1}(t) = \sigma_s^2 \int_{-\infty}^{\infty} h_1^2(\tau)\, d\tau \tag{7.20}$$

Considering both filters are identical, at the operational amplifier (OP-AMP), recall figure 7.6, two light intensities I_{S1} and I_{S2} will be subtracted, thus

$$I_{S1} - I_{S2} = \sigma_s^2 \int_{-\infty}^{\infty} [h_1^2(\tau) - h_1^2(\tau)]\, d\tau \tag{7.21}$$

This is what we had denoted by n_d in (7.13). Using the Parseval equality, we obtain

$$n_d = \sigma_s^2 \int_{-\infty}^{\infty} [|H_1(f)|^2 - |H_2(f)|^2]\, df = 0 \tag{7.22}$$

equation (7.22) indicates that the shot noise, which rises by the sunlight, in V2V-VLC can be neglected by using a differential detector.

7.5.3 Simulation results

We use Matlab to simulate the proposed system and consider the city of Toronto for calculating solar irradiance. Simulation parameters are given in tables 2.4, 4.1, 5.2, and 7.2 [7].

Figure 7.8 shows the response of the APD of the predicted spectral irradiance of the sunlight for different visible light wavelength. Here, APD response is given for three cases: without optical filter, with blue optical filter, and with differential receiver. Moreover, figure 7.8 shows that the violet and blue lights have less irradiance, while yellow and green lights have the highest irradiance. The differential receiver captured up to 60% of solar irradiance while the blue filter captured up to 75%, which means the proposed scheme shows good improvement in system performance.

The proposed system performance is studied by investigating the average BER versus the sunlight irradiance for different months of the year for the three scenarios, as shown in figure 7.9. A significant improvement is achieved when the differential receiver and blue filter are used. The differential receiver can improve the BER up to $\simeq 5 \times 10^{-5}$, while the blue filter scenario can improve the BER up to $\simeq 1 \times 10^{-4}$.

Figure 7.10 shows the data rate versus the sunlight irradiance for the three considered scenarios. It is shown that the proposed scheme improves the data rate three times comparing without any filtering scheme, and by one and a half times compared to a traditional filtering scheme. It is also shown by figures 7.9 and 7.10

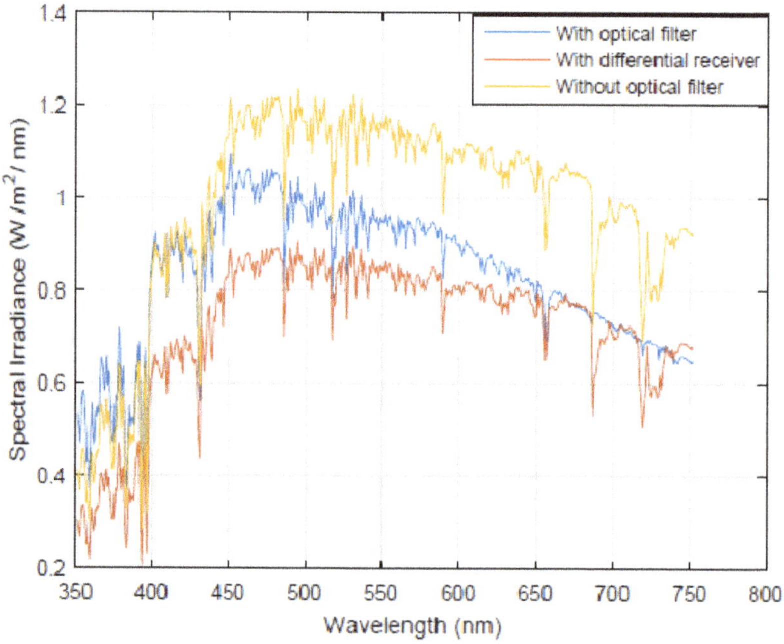

Figure 7.8. Average predicted solar irradiance at Toronto city in the month of June with the denoising schemes.

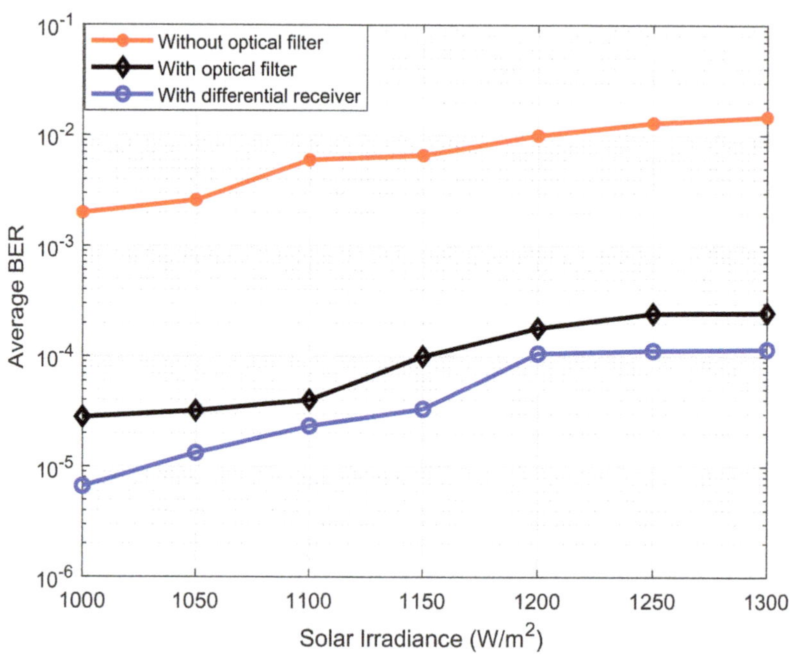

Figure 7.9. BER with sunlight irradiance considering the city of Toronto.

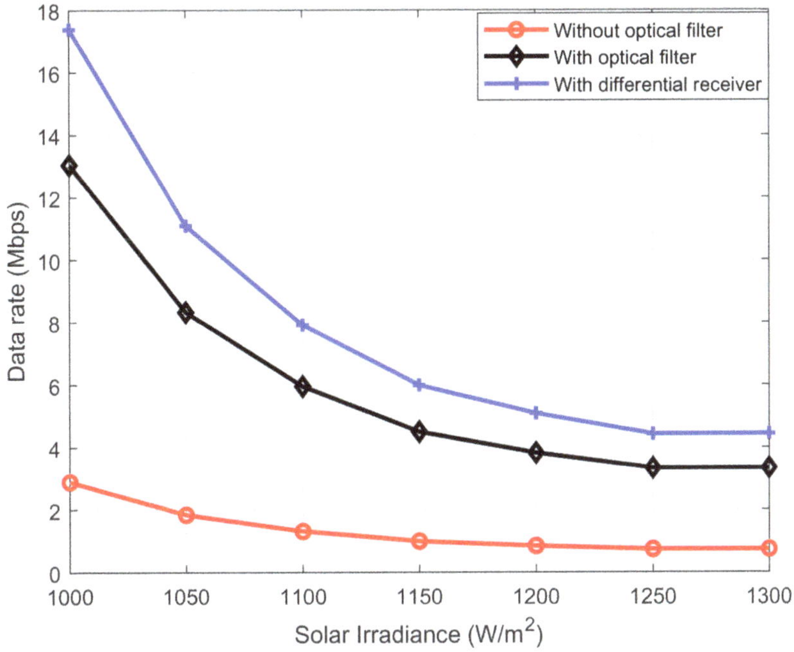

Figure 7.10. Achievable data rate with sunlight irradiance considering the city of Toronto.

that the V2V-VLC system can still work under solar irradiance, where the worst achievable BER during June (highest irradiance value) is $\simeq 5 \times 10^{-4}$.

7.6 Denoising of V2V-VLC systems using machine learning

ML is a powerful sub-field of artificial intelligence and is being used for model training purposes in data mining, pattern recognition, and image processing. Centralized and distributed algorithms are required for data fusion and automated decision making during smart sensing/monitoring. ML is used to develop technologies for machines/devices, for monitoring present behavior and predicting future behaviors. It has also been utilized in optical communication systems for indoor localization and failure prediction. Table 7.4 compares different ML approaches with their pros and cons.

ML techniques are becoming pervasive and they can play a key role in all stages of smart VLC networks—from data analysis to behavior prediction. In the second denoising scheme, we use *k*NN supervised learning technique, which is a simple and powerful algorithm intended to solve classification and clustering problems to combat the solar irradiance effect on V2V-VLC system. It is a straightforward and effective method to find and indicates associations within a given data set.

7.6.1 System model

In this study, we consider the same model which is given by section 4.3.2 and shown in figure 7.2 to apply ML for performance enhancement of VLC-based V2V systems

Table 7.4. Brief comparison of ML techniques for denoising V2X-VLC systems.

Technique	Advantages	Disadvantages
Standard K-nearest neighbors (*k*NN)	Simple and intuitive, non-parametric, effective for small datasets.	Computationally intensive for large datasets, sensitive to noise and irrelevant features. Requires large memory for storing training data.
Support vector machine (SVM)	Effective in high-dimensional spaces with clear margin of separation, robust to overfitting.	Not suitable for large datasets, Requires careful tuning of parameters and kernel selection, less effective with overlapping classes.
Random forest	Handles large datasets efficiently, reduces overfitting through ensemble learning, handles missing values.	Can be biased towards dominant classes,Lower interpretability, requires more computational resources.
Neural networks	Capable of learning complex patterns. highly scalable with large datasets. Can be fine-tuned for specific tasks.	Requires large amounts of data for training. Computationally intensive and requires specialized hardware. Prone to overfitting without proper regularization.
Principal component analysis (PCA)	Reduces dimensionality and noise. Enhances computational efficiency. Improves visualization of high-dimensional data.	Assumes linearity in data. Can lose important information in the process. Not suitable for non-Gaussian distributions.
Proposed scheme	Handles the problem at physical system level	Requires adjustment of transmitter irradiance and receiver FoV.

under a sunlight effect. Moreover, we have chosen the city of Toronto as a reference location for consideration of solar irradiance during the summer season. We consider the solar irradiance listed in table 7.2, and an OOK modulation scheme for VLC signal.

7.6.2 Machine learning and adaptive filtering

We apply an ML algorithm for the training of an optical filter. This filter is installed within the taillights (the receiver) of the receiving vehicle. This filter adaptively modifies itself responding to physical conditions of the VLC link—attempting to achieve an acceptable BER.

Depending upon the environmental conditions, the values for solar irradiance will change with weather and overtime day. The distance between two communicating vehicles will also vary due to the relative speeds between them. The FoV is expected to change with vehicle size and make, style and width of headlights and taillights. All of these factors effect the reception of the VLC signal and, in turn, effect the received BER.

We train our system using supervised learning. We use the nearest neighbour algorithm, which belongs to the 'supervised learning' ML class.

7.6.3 Why *k*NN algorithm

*k*NN algorithm is one of the simplest classification algorithms and it is one of the most used learning algorithms. Also, *k*NN is a non-parametric, lazy learning algorithm. Its purpose is to use a database in which the data points are separated into several classes to predict the classification of a new sample point. In our study, we use *k*NN because we are dealing with two variables only (FoV, and solar irradiance) so we do not need too much storage data nor high dimension data, thus, the calculation time will be reasonable. Also, we suggest a weighed voting factor to remedy skewed class distributions.

7.6.4 Problem classification

We train the system using the nearest neighbor supervised ML algorithm. We use data for the city of Toronto for various values of solar irradiance which is shown in table 7.2. After developing the data set for various parameters affecting the VLC signal, the nearest neighbor algorithm is applied to find the closest match of the actual transmitted VLC signal. The suggested algorithm finds the closest neighbor to the transmitted signal and finds the effective BER. If this BER is over a certain threshold, the suggested system modifies itself to achieve the desired acceptable rate.

7.6.5 Adaptive filtering

We do not give details of the filter design in this book and leave it for future work. There are many factors that can effect BER (transmitted power, incident angle, received angle, distance, solar irradiance), but for now, we consider only two parameters, i.e, FoV, and solar irradiance. We train our system using *k*NN, and it will adjust itself according to the difference between achieved and desired BER. The system can perform two actions, depending upon the received VLC signal and achieved BER.

- Blocking solar irradiance: We assume our filter is comprised of cascading films/blinds and can automatically make these films appear before APDs, for irradiance noise filtering from the VLC signal. The filter will adjust the number of films/blinds according to the intensity of solar irradiance. If the effect of solar irradiance is greater, the received BER is expected to be worse, therefore, the smart filter will automatically add more blinds to nullify the effect of ambient noise.
- Modifying the FoV angle: FoV can affect the VLC signal, as the incident angle of the received signal changes with different inclinations of taillight mounts. Changes in the incident angle will change the received signal strength and effect the BER. If achieved BER is not acceptable, our system will modify the inclination angle of taillights for proper FoV to achieve a better BER. The sequence of operations for our algorithm is given in figure 7.11.

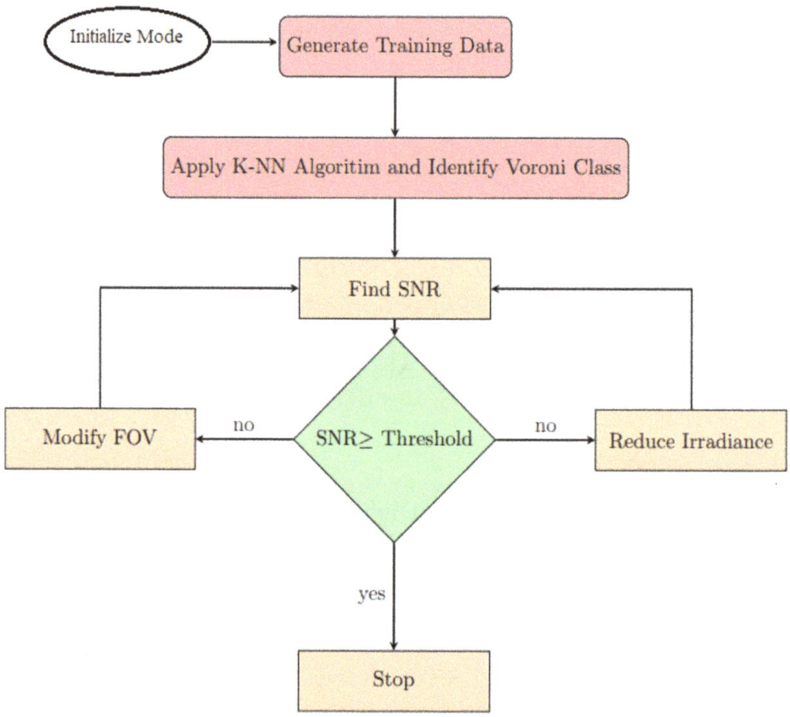

Figure 7.11. Adaptive filtering using *k*NN algorithm.

7.6.6 *k*-nearest neighbour algorithm and distance weighted *k*NN rule

*k*NN, is a well-known algorithm used in the pattern recognition literature. According to this algorithm, an unclassified pattern (sample, instance) is assigned to the class represented by a majority of its *k*-nearest neighbors. This rule is usually called the voting *k*NN rule. The number N of patterns and k are related such that, $k/N \mapsto 0$, the error rate of the *k*NN rule approaches the optimal Bayes error rate.

*k*NN regression is a non-parametric regression method, where the information derived from the observed data is applied to forecast the number of predicted variables in real time without defining a predetermined parametric relation between predictor and predicted variables. Also, in voting *k*NN, the k neighbors are implicitly assumed to have equal weight in the decision, regardless of their distances to the observed data x to be classified, or, otherwise, the weight should be define for each neighbor. The basis of this method is on calculating the similarity (neighborhood) of the real-time number of predictors $X_r = x_{1r}, x_{2r}, x_{3r}, \dots, x_{mr}$ with the amount of predictors for each historical observation $X_t = x_{1t}, x_{2t}, x_{3t}, \dots, x_{mt}$ via Euclidean distance function D_{Ec} or any other distance method. In this study, we use two methods; Euclidean distance and city block method (Manhattan distance) in seeking comparison of performances.

Euclidean distance is given as follows

$$D_{Ec} = \sqrt{\sum_{i=1}^{m} w_i\,(x_{ir} - x_{it})^2} \qquad t = 1, 2, 3, \ldots, n \tag{7.23}$$

where $w_i(i = 1, 2, \ldots, m)$ are the weights of the predictors, summation of which is equal to one.

The city block distance between two points is calculated as

$$E_{bo} = \sum_{i=1}^{K} |(x_{ir} - x_{it})| \tag{7.24}$$

In this study, the weight w_i is calculated as follows

$$w_i = \begin{cases} \dfrac{d(x_k, x) - d(x_i, x)}{d(x_k, x) - d(x_1, x)} & \text{if} \quad d(x_k, x) \neq d(x_1, x) \\ 1 & \text{if} \quad d(x_k, x) = d(x_1, x) \end{cases} \tag{7.25}$$

where $d(x_k, x)$ is the distance between the kth neighbor and the observation x.

More formally, given a positive integer k, an unseen observation x and a similarity metric distance, kNN classifier performs the following two steps:

Step A: distance between data points: For certain values of k, the distance between observed data and its neighbors can be calculated using different methods such as Minkowsky, Euclidean, city block, etc. Afterward, measured distances are sorted to determine the nearest neighbor based on the kth minimum distance. Thereafter, the categories of the nearest neighbors are gathered to do the voting. The algorithm runs through the entire data set, computing the distance D between x and each training data element. k points in training data closest to x are saved in set A.

Step B: categorial assignment: For an instance x (observed data), assume we have N_i instances belong to class Y_i in the neighborhood. Then we define

$$P(Y_i/x) = \frac{N_i + s}{k + Cs} \tag{7.26}$$

where k is the total number of instances in the neighborhood, C is the total number of classes, and s is the smoothing parameter. The smoothing is used to avoid 0 probabilities.

A detailed self-explanatory description of our proposed algorithm is given in algorithm 1.

Algorithm 1: k-nearest neighbor.

Classify (Xi, Ci, x); where $i = 1, 2, 3, \ldots, k$ are the data points, X: feature values, C: class labels of Xi for each i, x is the unknown sample

BEGIN

for $i = 1$ to $i = K$ do

Compute distance $d(Xi, x)$, where d denotes the distance between the two points

end for
Compute set A, containing indices for K smallest points $d(Xi, x)$
return majority label for Ci **where** $i \in A$.

7.6.7 Simulation results

In the simulation, we use the ML toolbox in the Matlab suit of software. The simulation parameters are shown in tables 2.4, 4.1, 5.2, and 7.2.

7.6.7.1 Data sets and classes

In figure 7.12, a Voronoi diagram is shown to represent the training and the observed data and their distribution to various classes on the basis of calculated BER. We plot datasets as a Voronoi diagram to make it easy to recognize the boundaries of decisions areas. The classification of a data set is based on the classes of its nearest neighbors. Since each data set consists of two variables; solar irradiance, and FoV, the *x*-axis represents the irradiance and the *y*-axis represents the FoV.

The functionality of the *k*NN algorithm can be explained by considering one element of the data set. We denote the observed data in Voronoi, with a small circle 'o', and the training data by small '*', for a certain observed data point. For a certain value of *k*, the distance between observed data and its neighbors can be calculated. Afterward, measured distances are sorted to determine the nearest neighbor based on the *k*th minimum distance.

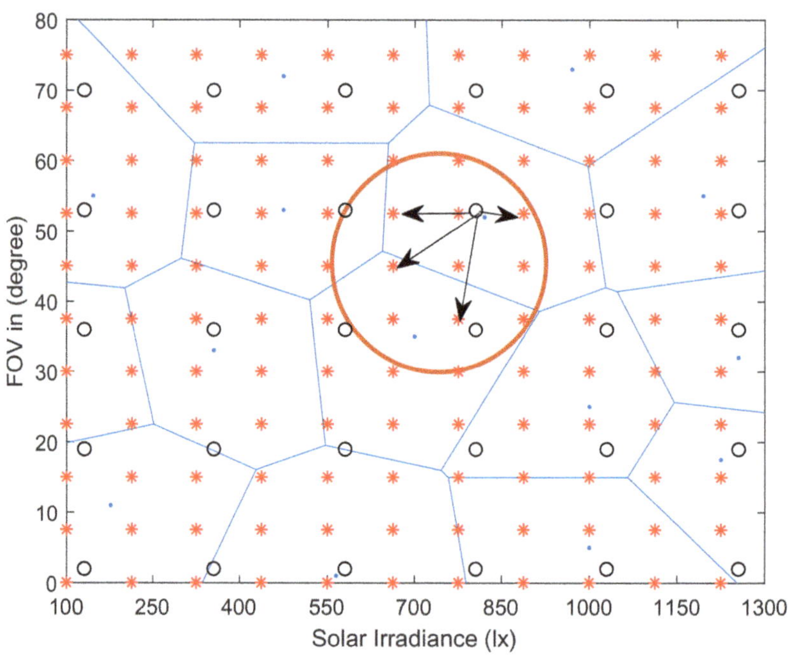

Figure 7.12. Training data and observed data in a Voronoi diagram.

7.6.7.2 Choice of k

k in *k*NN is the number of instances that we take into account for the determination of affinity with classes. So we need to investigate the performance of *k*NN near rule-of-thumb-value and make a decision about the optimal one using any algorithm for performance testing. In figure 7.13, the effect of the numbers of selected neighbors on the decision accuracy is presented. We use the Euclidean distance formula and city block distance formula to generate this figure, and use a Monte Carlo method to average it over 2000 iterations. It is shown that optimal value of *k* is 12 with Euclidean distance and *k* is 6 with city block distance.

7.6.7.3 BER discussion

Figure 7.14 represents the BER of the proposed system before applying the denoising scheme and after applying the scheme. Figure 7.14 shows improvement in terms of BER after applying the denoising scheme. We can notice an improvement in BER from 10^{-2} to 10^{-5}, thus, reflecting the stability and quality of the communication link.

Figure 7.15 shows the probability of error of each action (add optical filter or change the FoV) at each daytime hour. Because choosing the action is random, this result may help to determine which action of the system has to be chosen at each hour of the day. Moreover, figure 7.15 shows the probability of adding an optical filter is higher than changing the FoV, the reason is that while times change, the position of the Sun also changes and the irradiance value will change.

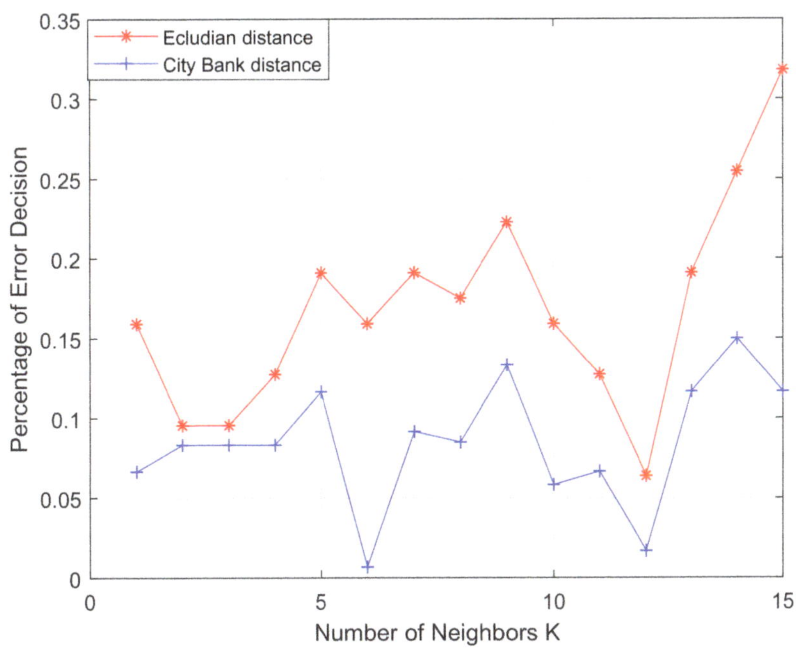

Figure 7.13. The relation between the number of neighbors and the percentage of decision error.

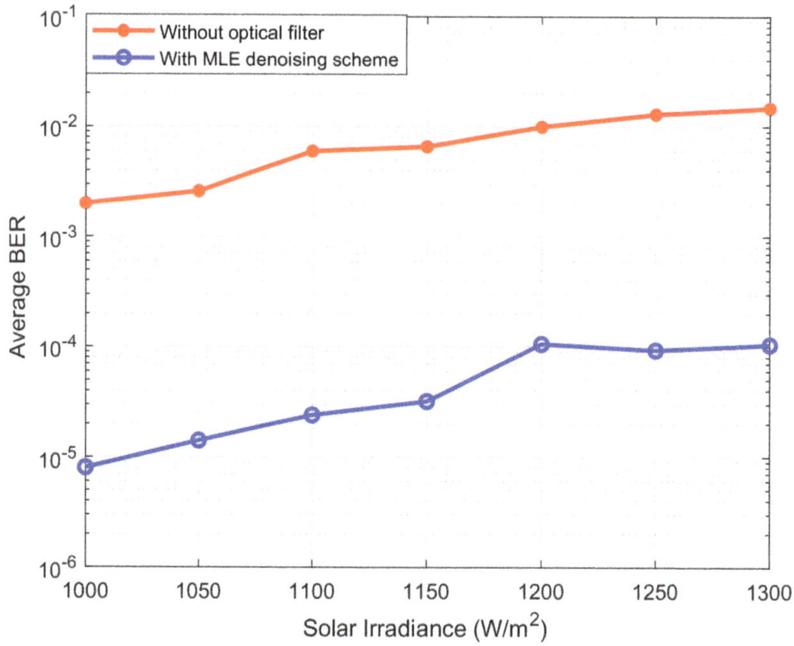

Figure 7.14. BER of the V2V-VLC system with and without denoising scheme.

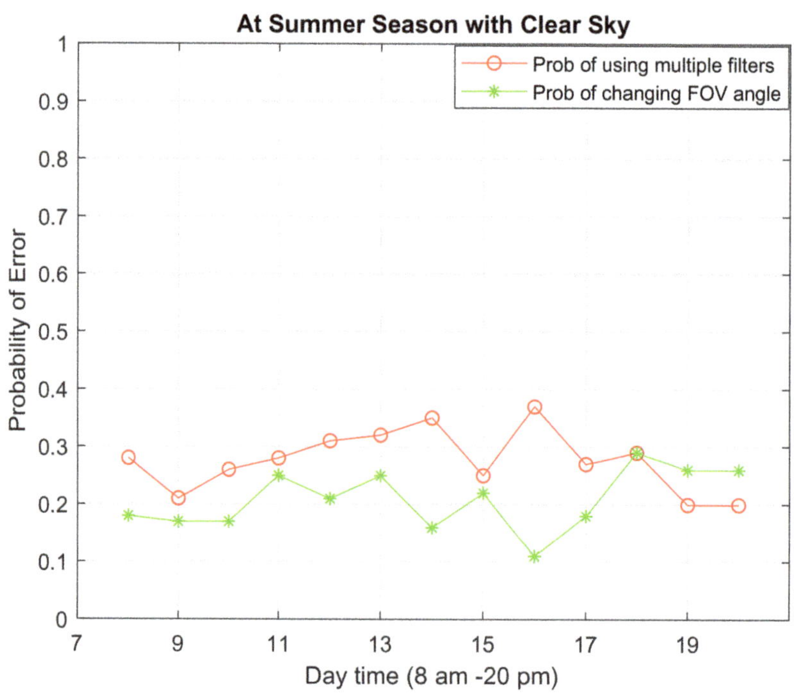

Figure 7.15. Probability of each action with day hours.

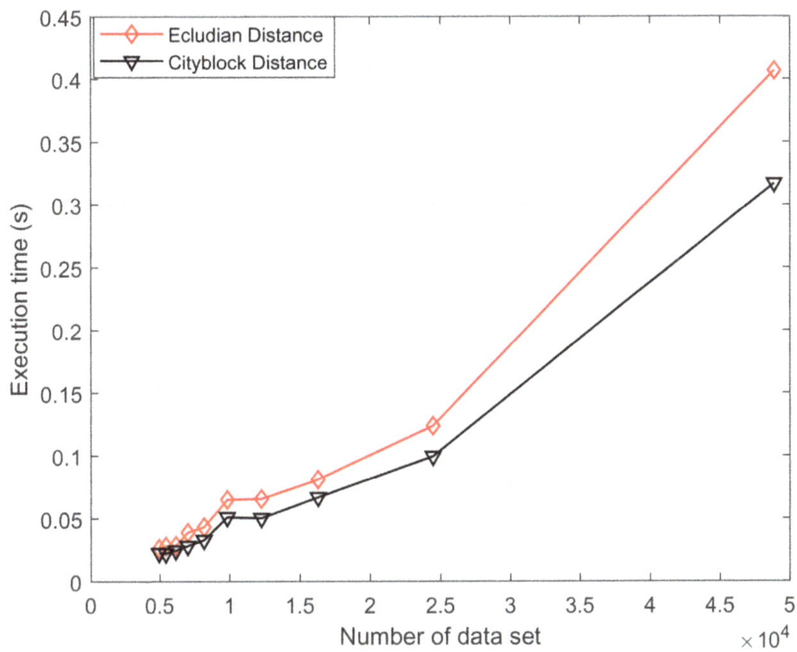

Figure 7.16. Execution time for the proposed scheme.

7.6.8 Execution time

In figure 7.16, we discuss the execution time for the proposed scheme. We consider two methods to calculate the distance between the instance and its neighbors, Euclidean distance, and city block distance. Also, we used an Intel(R) Core (TM) i5-5200U CPU@ 2.20 GHz laptop to calculate the execution time for the datasets. Figure 7.16 shows both methods almost have the same time of execution for low numbers of datasets 0.05 s for 5000 datasets, while the city block method is better for a high number of datasets 0.32 s for 50 000 datasets.

7.7 Chapter summary

This chapter proposes two denoising schemes to combat the effect of solar irradiance, thus, improve the performance of V2V-VLC systems under sunlight effect. The first proposed denoising scheme is by using a differential receiver, while the second proposed scheme is by using a *k*NN algorithm as an ML-based adaptive filter. Simulation results of both schemes show good improvement in SNR, which makes the communication link more stable, even in summer months.

References

[1] Solar Power Calculator for Toronto, Ontario, Canada https://solarcalculator.ca/report/Ontario/Toronto/ (Accessed: Dec 10, 2022)

[2] Photodiode parameters https://www.hamamatsu.com/jp/en.html (Accessed: June 28, 2023)

[3] Solar calculations www.solarelectricityhandbook.com/index.html (Accessed: Dec 10, 2022)

[4] Chung Y H and Oh S B 2013 Efficient optical filtering for outdoor visible light communications in the presence of sunlight or articifical light *2013 Int. Symp. on Intelligent Signal Processing and Communication Systems* pp 749–52
[5] Deng Z, Jiang Y, Ma Z, Wang W, Jia H, Zhou J and Chen H 2013 A novel wavelength-adjusting method in InGaN-based light-emitting diodes *Sci. Rep.* **3** 3389
[6] Duffie J A and Beckman W A 2013 *Solar Engineering of Thermal Processes* (New York: Wiley)
[7] Fernando X, Farahneh H and Kamruzzaman S 2018 Differential receiver as a denoising scheme to improve the performance of V2V-VLC systems *ICC 2018-USA* vol 10 pp 45–53
[8] Islim M S and Haas H 2017 An investigation of the sunlight irradiance effect on visible light communications *2017 IEEE 28th Annual Int. Symp. on Personal, Indoor, and Mobile Radio Communications (PIMRC)*
[9] Karaboga N 2005 Digital IIR filter design using differential evolution algorithm *EURASIP J. Adv. Signal Process.* **2005** 1269–76
[10] Keiser G 2003 *Optical Communications Essentials* (New York: McGraw-Hill)
[11] Khaligh M-A, Xu F, Jaafar Y and Bourennane S 2011 Double-laser differential signaling for reducing the effect of background radiation in free-space optical systems *IEEE/OSA J. Opt. Commun. Netw.* **3** 145–54
[12] Khan M F F, Akif A and Haque M A 2017 Iris recognition using machine learning from smartphone captured images in visible light *2017 IEEE Int. Conf. on Telecommunications and Photonics (ICTP)* pp 33–7
[13] Komine T and Nakagawa M 2004 Fundamental analysis for visible-light communication system using LED lights *IEEE Trans. Cons. Electron.* **50** 100–7
[14] Lee I E, Sim M L and Kung F W L 2009 Performance enhancement of outdoor visible-light communication system using selective combining receiver *IET Optoelectron.* **3** 30–9
[15] Lee I E, Sim M L and Kung F W L 2011 A dual-receiving visible-light communication system under time-variant non-clear sky channel for intelligent transportation system In *2011 16th European Conf. on Networks and Optical Communications (NOC)* pp 153–6
[16] Andonovic I, Beshr M and Hussien M 2012 The impact of sunlight on the performance of visible light communication systems over the year *Proc. of SPIE-The Int. Society for Optical Engineering* pp 1387–91
[17] Sun W, Zhao C, Chen L, Li D, Bai Y, Jia W and Sun M 2015 Learning based particle filtering object tracking for visible-light systems *Int. J. Light Electron Opt.* **126** 1830–7
[18] Guo X, Shao S, Ansari N and Khreishah A 2017 Indoor localization using visible light via fusion of multiple classifiers *IEEE Photon. J.* **9** 1830–7

Xavier Fernando and Hasan Farahneh

Chapter 8

Hybrid channel-based foglet-assisted smart asset reporting

There is a lack of focus on smart asset management in abandoned areas and at the sides of roads. The majority of labor is found in emergency reporting services in rural and intelligent transportation system (ITS) locations; asset reporting is not as prevalent. The asset management systems in place today rely on planned maintenance and are ineffective at promptly reporting emergencies. In this chapter, we present an architecture for smart transportation for asset management and emergency reporting by using a hybrid communication system. Here we will use foglets with radio frequency (RF) spectrum (wireless) and visible light communication (VLC). We introduce the hybrid model's parameters and channel model to facilitate the intelligent transportation system (ITS). Simulations demonstrate a significant increase in the system's performance with regard to communication range and emergency reporting time.

8.1 Introduction

A hybrid free-space optical/millimeter wavelength (MMW) which is a part of an RF link can be applied to enhance performance of free-space optical (FSO) communications. This is due to the wide variations in how weather and atmospheric conditions affect FSO and RF links.

FSO lines are less impacted by rain but suffer from exceptionally high attenuation when there is fog. By contrast, rain greatly increases link attenuation, whereas fog essentially has no effect on MMW RF lines. Similar to how atmospheric turbulence mostly causes small-scale fading in FSO systems, multipath propagation causes fading in RF networks. In addition to having high data speeds that are equivalent to FSO links, MMW RF links provide intrinsic security because of high link attenuation, flexible deployment options, and license-free operation [1].

Because FSO and MMW RF lines complement each other, there are a number of ways to create hybrid RF/FSO data transmission systems. There have been two

doi:10.1088/978-0-7503-6049-4ch8

primary methods for putting hybrid RF/FSO systems into practice. The switch-over hybrid RF/FSO technique is one method that uses hardware switching to transition between FSO and MMW RF links. However, this approach will lead to frequent hardware switching between the FSO and RF lines.

Using RF and FSO lines simultaneously for data transfer is an additional strategy. Using diversity combing techniques on the received signals from both lines, one method in this approach is to transmit identical data concurrently on both links. Thus, the data rate of the system is restricted to the lower rate of the RF link [2].

Splitting the data stream with codes between the two links is an additional method that could significantly increase system capacity. Generally speaking, this soft-switching technique necessitates that FSO and RF links remain active at all times, even in cases when the FSO link is of excellent quality and can sustain the necessary bit-error rate on its own. In this case, the system wastes RF transmission power and causes needless RF interference to the surrounding area.

We were inspired to create a novel hybrid RF/FSO transmission scheme (figure 8.1) by these shortcomings of the hybrid RF/FSO systems that had previously been demonstrated. Hybrid RF/FSO transmission with adaptive combining is the name given to this technique [3]. As long as the data transmission quality is satisfactory and the RF connection is placed in standby mode, this technique uses the FSO link exclusively for data transmission.

When the FSO link's quality declines to an unacceptable level, the system turns on the RF link and processes signals coming in from the FSO and RF lines using the maximal Ratio Combining (MRC) approach. Moreover, when the quality of the FSO link alone is again judged sufficient, the RF link is again put in standby mode to save power and use less spectrum.

Therefore, the adaptive combining scheme that has been proposed for hybrid RF/FSO systems for the following reasons:

- Increases the dependability of the communication system by preventing problems with switch-over schemes.
- Avoids creating needless radio frequency interference to the environment.

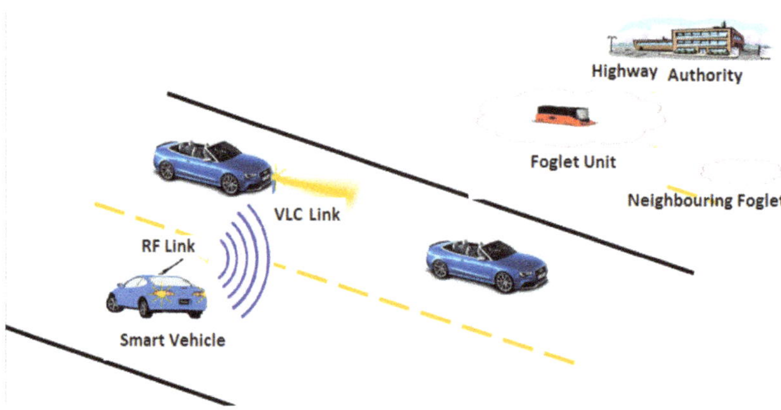

Figure 8.1. Block diagram of the proposed system.

- Maintains RF power.
- The majority of the time it gains from FSO's increased data rate.

The rest of this chapter is structured as follows: we describe our model in section 8.2. The performance analysis of the suggested system is presented in section 8.3. Section 8.4 presents the analysis and findings of the simulation. Section 8.5 contains the conclusions.

8.2 System model

As seen in figure 8.2, our suggested system combines RF and VLC communication. One essential aspect of optical communication is good directionality. However, the existence of impediments causes the optical link to break, and the optical channel is affected by ambient light interference and weather. Radio waves, on the other hand, are not very directed but they are very reliable.

Users of foglets frequently need higher directional communication, such as when they need information on a single lane. Figure 8.2 displays a hybrid VLC/RF communication system for the downlink that spans a wide region. The system is made up of VLC links, and each VLC link is made up of an LED lamp with many LEDs, such as the headlights or taillights of a car. The VLC link can only cover a limited area over short distances because we are just taking into account line-of-sight (LoS) transmission.

To increase communication coverage and link availability, an RF communication system is added to the basic VLC system in this study. For this system, we suggest a new joint adaptive combining and power adaptation approach. Under this technique, data transmission occurs only over a high-speed VLC link until the signal quality drops to a level that is acceptable. The system initiates the RF link when the quality of the VLC link is no longer adequate, allowing the identical data to be transferred simultaneously over the two links.

The signals received from both lines are then subjected to maximal ratio combining (MRC) by the receiving terminal. In this instance, and to match the RF link and to aid with diversity reception, the system reduces the VLC link's transmission rate.

Moreover, when the radio frequency link is activated, and to adjust the transmit power over the link, a modified truncated channel-to-inversion (TCI) power adaptation policy is applied. Thus, in order to achieve an acceptable system

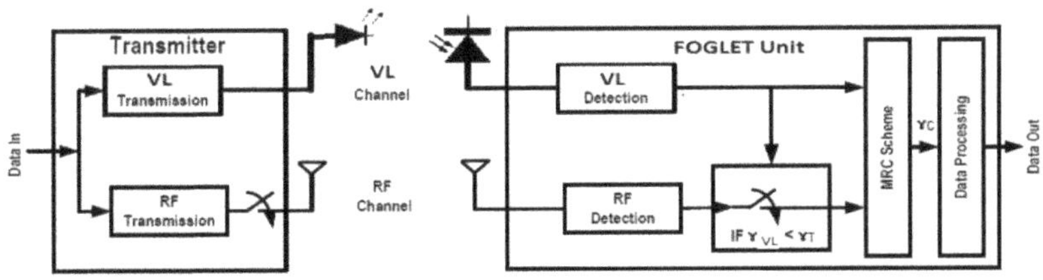

Figure 8.2. Hybrid VLC/RF assisted asset management infrastructure.

performance, the MRC combination of the RF and VLC links keeps the received signal-to-noise ratio (SNR) constant. Improving link dependability and average user throughput is the goal here.

Since omnidirectional transmit antennas make up RF networks, a broad coverage area is anticipated. There is a central unit (CU) that manages the RF and VLC systems. Based on predetermined performance standards, the CU dynamically chooses the appropriate connection to serve the specific user, making the VLC system the primary and the RF system the secondary system that serves as a stand-in for the VLC system. The absence of interference between RF and VLC channels is a major advantage of implementing VLC in future heterogeneous networks. Performance is improved as a result, and VLC networks offer fast data rates free from interference.

8.2.1 VLC channel between the vehicle and the foglets

In our setup, the foglet serves as the receiver for the VLC link and the vehicle acts as a transmitter. We simply look at the LoS path. Additionally, we take into account the headlamps of a market-weighted car with a lambertion source illumination pattern [4]. Unfortunately, the main challenge of the VLC system is the background noise.

Since the photodiode (PD) is subjected to both necessary optical signals and other artificial or natural lighting at the receiver side, this is regarded as undesirable optical noise interference. The noise involved in an optical wireless channel can be modeled as given in equation (4.1) and as shown in [5].

Here, we consider the direct path between the transmitter and the foglet.

For the LoS path, the impulse response between the transmitter T and receiver R at the foglet station, $h_{RT_{\mathrm{LOS}}}$ is given by equation (4.3) and shown in [6].

The received power for the LoS is given by equation (4.42) and shown in [7].

We use the on–off keying (OOK) modulation technique and a Gaussian channel in this work. We must determine the BER and noise power in order to examine the functionality of our suggested solution. During the day, the primary source of noise for VLC links is still shot noise caused by solar radiation because interference from artificial lights is less intense than that of solar radiation. Nonetheless, the primary cause of noise at night is artificial light interference [8]. We used OOK modulation and a Gaussian noise with a total variance equal to the sum of the contributions from the thermal noise and shot noise associated with the receiver in our suggested system.

It is significant to remember that, in contrast to the few MHz of bandwidth utilized in a VLC system, the usual channel latency for a VLC link is 10 ns. As a result [9], ISI brought about by the multipath environment is minimal.

Thus, at the receiver, the electrical SNR is given by:

$$SNR = \frac{(\gamma\ P_R)^2}{\sigma^2_{\mathrm{total}}} \tag{8.1}$$

the total noise variance σ^2_{total} is given by:

$$\sigma^2_{\text{total}} = \sigma^2{}_{\text{shot}} + \sigma^2{}_{\text{thermal}} \tag{8.2}$$

and the BER is given as:

$$BER_{op} = \frac{1}{2}\, erfc\left(\frac{1}{2\sqrt{2}}\sqrt{\frac{(\gamma P_r)^2}{\sigma^2_{\text{total}}}}\right) \tag{8.3}$$

where the complementary error function, *erfc*(.), is used to get the conventional Gaussian distribution's tail probability.

As a result, the data rate C that the user can achieve while fully utilizing the available bandwidth is as follows:

$$C = B\,\log_2(1 + SNR) \qquad bps \tag{8.4}$$

where, B is the system bandwidth.

Since the VLC system only covers a limited area, interference from other light sources' visible light can be rather strong.

In a traditional pure VLC system, the data stream is transmitted via the optical signal; a code signal is not required. The VLC system's detection error probability as reported by [10]

$$P_{\text{Error}} = 1 - (1 - BER_{op})^{L_{\text{DATA}}} \tag{8.5}$$

where L_{DATA} is the data frame's information bit count.

The quantity of data obtained in a service region that can offer trustworthy communication is expressed as [10]:

$$M_p = R_b\, L_{SA}/v \tag{8.6}$$

where the length of the service area is L_{SA}, the car speed is v, the transmission bit rate is R_b, and the receivable information is M_p.

8.2.2 Radio frequency link

The information signal is transmitted via the RF channel. In this case, we'll assume that the RF signal is transmitting across the LoS channel and that its transmit power is P_{RFt}.

The RF signal's received power can be written as [10]:

$$P_{\text{RF}s} = P_{\text{RF}t}\left(\frac{\sqrt{G_l}\, c/f_{\text{RF}}}{4\pi\, d_{\text{RF}}}\right) \tag{8.7}$$

where, the product of the transmit and receive antenna field radiation pattern is $\sqrt{G_l}$, and c is the speed of light, d_{RF} is the distance between the RF transmitter and receiver at the foglet station, and f_{RF} is the radio frequency, which is assumed to be 2.4 GHz.

It is good to mention that we consider the antenna is omnidirectional.

Then, the SNR of the radio channel is:

$$SNR_{\mathrm{RF}} = \frac{P_{\mathrm{RF}s}}{B_{\mathrm{RF}}\,\sigma^2_{\mathrm{thermal}}} \tag{8.8}$$

where B_{RF} is channel bandwidth.

The signal-to-interference-plus noise ratio (SINR) of radio channel for K users can be given as [11]:

$$SINR_{\mathrm{RF}} = \left(\frac{B_{\mathrm{RF}}\sigma^2_{\mathrm{Thermal}}}{P_{\mathrm{RF}s}} + \frac{K-1}{G_{\mathrm{RF}}}\right)^{-1} \tag{8.9}$$

where G_{RF} is the processing gain.

In this chapter, When using binary phase shift keying (BPSK) modulation, the radio system's BER is given as [11]:

$$BER_{\mathrm{RF}} = Q(\sqrt{2\,SINR_{\mathrm{RF}}}) \tag{8.10}$$

8.3 Performance analysis

In a traditional VLC system, data transmissions are transmitted over short distances using optical signals and over long distances using RF links.

We examine a hybrid VLC/RF system that combines power adaptability and adaptive combining. The VLC link at the car is taken to be always operational and to have continuous transmit power in this hybrid VLC/RF system. In order to enable simultaneous transmission of the same data over both the RF and VLC links, the foglet station provides a feedback signal when the instantaneous SNR at the VLC link's receiver γ_{VLC} falls below the threshold γ_T. We can presume that instantaneous channel state information (CSI) of both links is always available at the receiver due to the channels' slowly time-varying nature, and that there is a feedback channel for CSI feedback.

An MRC combiner at the foglet station will combine the data received over the two lines. The receiver SNR γ_C will equal γ_{VLC}, when $\gamma_{\mathrm{VLC}} \geqslant \gamma_T$. When $\gamma_{\mathrm{VLC}} < \gamma_T$, γ_C will equal the sum of γ_{VLC} and γ_{RF}.

In this instance, the RF link power transmission is adjusted to maintain a consistent received SNR of γ_C i.e., to keep $\gamma_{\mathrm{VLC}} + \gamma_{\mathrm{RF}} = \gamma_T$.

The suggested joint method states that when employing a VLC link, the hybrid VLC/RF system never experiences an outage condition as long as it is able to maintain the target received SNR γ_T. This is because the system initiates the RF link for the MRC combination of the received SNRs on both VLC and RF links when γ_{VLC} is less than γ_T.

Only when the received SNR γ_{RF} of the RF link falls below γ_0 does the hybrid system enter an outage state. This case's outage probability can be computed using the following formula:

$$\begin{aligned} P_{\text{out}} &= P_r[\gamma_{\text{VLC}} < \gamma_T]\, P_r[\gamma_{\text{RF}} < \gamma_0] \\ &= F_{\gamma_{\text{VLC}}}(\gamma_T)\, F_{\gamma_{\text{RF}}}(\gamma_0) \end{aligned} \tag{8.11}$$

where $F_{\gamma_{\text{VLC}}}(.)$ and $F_{\gamma_{\text{RF}}}(.)$ are the cumulative distribution functions (CDF) of γ_{VLC} and γ_{RF} where those functions are given by [12].

To evaluate the outage capacity for the proposed hybrid system, we assume slowly-varying channels, which are the case with VLC and MMW RF channels with bit-duration much smaller than the channels' coherence time.

The hybrid VLC/RF communication system's outage capacity, normalized to the communication channel bandwidth B, can be written as follows for a given outage probability P_{out} [13].

$$\frac{C(P_{\text{out}})}{B} = [1 - P_{\text{out}}]\log_2(1 + K\,\gamma_T) \tag{8.12}$$

where K is a constant term such that $K = 1$ for heterodyne detection giving an exact result and $K = \frac{e}{2\pi}$ for intensity modulation/direct detection.

8.4 Results and discussion

We conducted the simulation for our proposed system using Matlab program. Table 8.1 displays the parameters used in the simulation.

The received data at various bit rates, assuming no roadblocks, is displayed in figure 8.3. It is evident that the suggested system has a higher bit rate. In actuality, an optical link's continuous transmission time may be extremely brief in congested traffic situations. The data needs to be sent again when the optical link is broken.

As the transmitted power grows, the hybrid VLC/RF system's outage performance improves, as illustrated in figure 8.4. Also, BER for the hybrid system hit the target BER (10^{-5}) for a reasonable transmitted power which mean saving in energy could be one benefit of the hybrid system.

Table 8.1. Comparison of VLC AND RF (DSRC) schemes.

Parameter	Value
Transmitted optical power	25 W
BW of LED, B	40 MHz
The receiver (PD) area	1 cm^2
Receiver FoV	85°
Receiver resistivity γ	0.53 A W^{-1}
Power of the noise	10^{-21}
Carrier frequency	2.4 GHz
Antenna gain at TX	43 dBi
Antenna gain at RX	43 dBi
Link distance	1.5 km

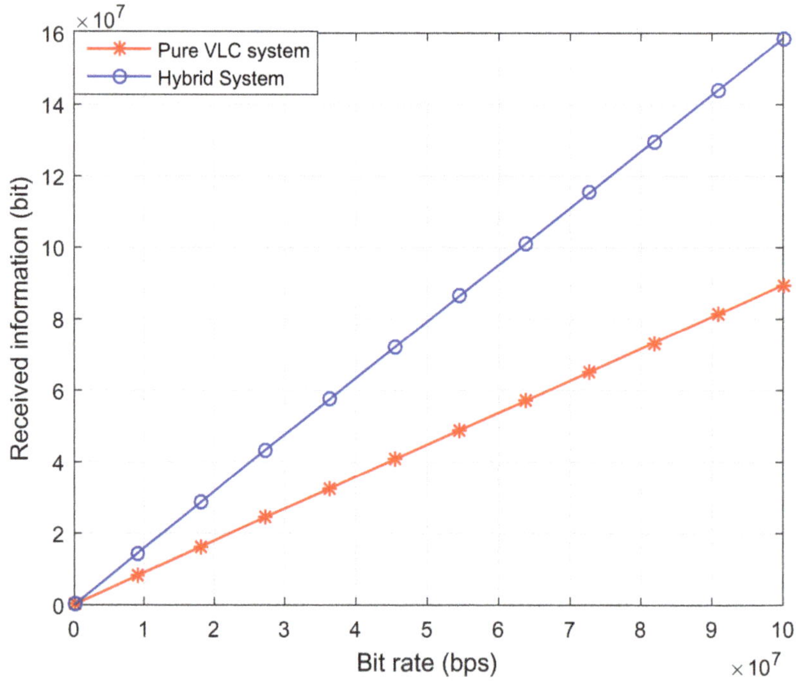

Figure 8.3. Receivable information rate versus bit rate.

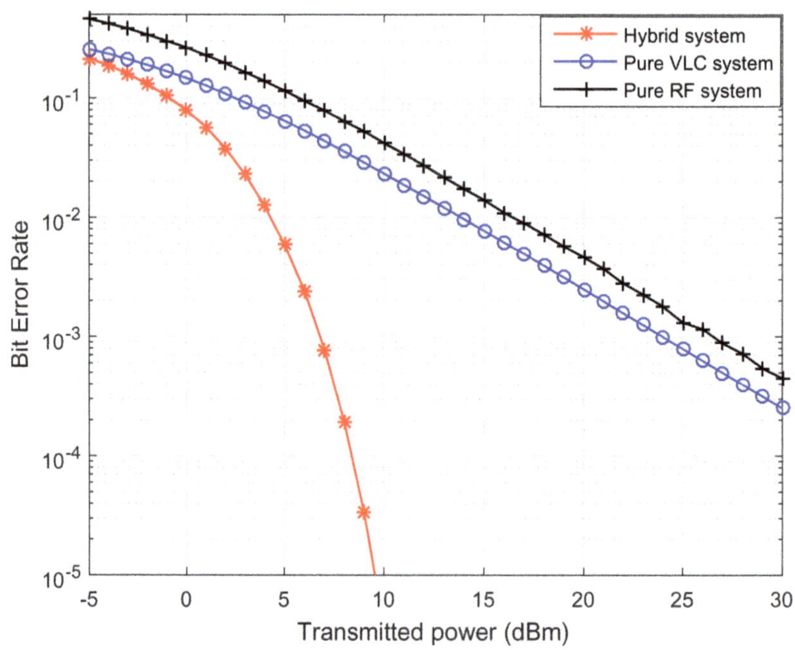

Figure 8.4. The BER of a hybrid VLC/RF system.

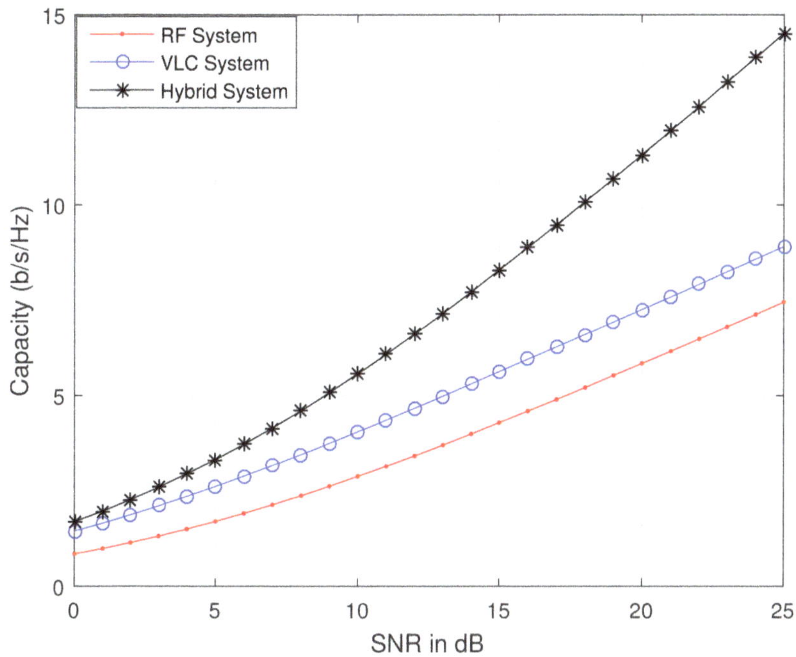

Figure 8.5. The capacity of a hybrid VLC/RF system.

Figure 8.5 shows the relation of the SNR of the pure VLC system, pure RF system, and the hybrid system where a notable improvement can be detected in the capacity of the system channel.

8.5 Chapter summary

In order to improve ITS and increase the downlink broadcasting efficiency of the current traffic system, this chapter proposed a hybrid VLC/RF foglet system. When the transmitter and receiver are close enough to each other to be considered as part of a VLC channel, the proposed system can use it to send data to the foglet; otherwise, an RF link is used. Both BER and channel capacity exhibit appreciable improvements in the hybrid system. Additionally, it enhances the transmitted power decreases.

References

[1] Odeyemi K O and Owolawi P A 2019 Selection combining hybrid FSO/RF systems over generalized induced-fading channels *Opt. Commun.* **433** 159–67

[2] Feng J and Zhao X 2016 Performance analysis of mixed RF/FSO systems with STBC users *Opt. Commun.* **381** 244–52

[3] Vishwakarma N 2021 Performance analysis of hybrid FSO/RF communication over generalized fading models *Opt. Commun.* **487** 126796

[4] Paudel *et al* 2013 Modelling of free space optical link for ground-to-train communications using a Gaussian source *Optoelectronics IET* **7** 1–8

[5] Kahn J B J M 1997 Wireless infrared communications *Proc. IEEE* **85** 265–98
[6] Lee, Sim M L and Kung F W L 2009 Performance enhancement of outdoor visible-light communication system using selective combining receiver *IET Optoelectron.* **3** 30–9
[7] Luo P, Ghassemlooy Z, Minh H L, Bentley E, Burton A and Tang X 2014 Fundamental analysis of a car to car visible light communication system *2014 9th Int. Symp. on Communication Systems, Networks Digital Signal Processing (CSNDSP)* pp 1011–6
[8] Hongsong C, Zhongchuan F and Dongyan Z 2011 Security and trust research in M2M system *IEEE Vehicular Technology Conf.* pp 286–90
[9] Lee S J *et al* 2012 Evaluation of visible light communication channel delay profiles for automotive applications *EURASIP J. Wirel. Comm. Netw.* 1823–6
[10] Liu J, Chan P W C, Ng D W K, Lo E S and Shimamoto S 2012 Hybrid visible light communications in intelligent transportation systems with position based services in *2012 IEEE Globecom Workshops* pp 1254–9
[11] Goldsmith A 2005 *Wireless Communications* (Cambridge: Cambridge University Press)
[12] Farid A A and Hranilovic S 2007 Outage capacity optimization for free-space optical links with pointing errors *J. Lightw. Technol.* **25** 1702–10
[13] Wang X and Liang Q 2012 On the outage throughput capacity of hybrid wireless networks over fading channels *2012 IEEE Global Communications Conf. (GLOBECOM)* pp 2173–8

www.ingramcontent.com/pod-product-compliance
Ingram Content Group UK Ltd.
Pitfield, Milton Keynes, MK11 3LW, UK
UKHW051852100926
8556IPUK00005B/584

9 780750 360470